THIS BOOK IS CRUELTY FREE

First published in the UK in 2021 by
Pavilion Books Company Limited
43 Great Ormond Street
London, WC1N 3HZ

Image credits:
lizard by Serhii Smirnov from the Noun Project
Flying Parrot by Agne Alesiute from the Noun Project
Spider by Miroslava from the Noun Project
monkey by TkBt from the Noun Project

Publisher: Neil Dunnicliffe

Editorial Reviewer: Alice Corrie

Designer: Sarah Crookes

Illustrations: Josephine Skapare

ISBN: 9781843654902

A CIP catalogue record for this book is available from the British Library.

10 9 8 7 6 5 4 3 2 1

Reproduction by Mission, Hong Kong
Printed and bound by Bell & Bain Limited, UK

This book can be ordered at www.pavilionbooks.com, or try your local bookshop.

THIS BOOK IS CRUELTY FREE

ANIMALS AND US

Linda Newbery

PAVILION

CONTENTS

“

IT’S NOT DIFFICULT TO REALIZE THAT ANIMALS HAVE FEELINGS. LIKE US, THEY EXPERIENCE FEAR, BOREDOM, ANTICIPATION AND PLEASURE.

INTRO

If you have a dog at home - or if you spend time with a friend or relative's dog - you'll know that it has different moods, behaviours, likes and dislikes. It lets you know very clearly when it wants to play or go for a walk, or when it's bored, frightened or hungry.

I've never owned a dog, but I've had a number of cats over the years, most of them re-homed from animal shelters. At present I have two cats, plus a regular visitor - a neighbour's cat who calls in most days. All three have distinct personalities: one of ours, Fleur, is talkative and bossy. The other, Holly, is placid and gentle, while visitor Louis is lazy and sometimes playfully aggressive.

Anyone who's kept chickens knows that they have favourite places and food treats, and they love to sunbathe, spreading their wings. They make a range of sounds: alarm calls, squeaks of excitement when given something tasty to eat, and the crooning purr that means they're really content.

It's not difficult to realize that animals have feelings. Like us, they experience fear, boredom, anticipation and pleasure. I think few people would deny that.

Yet the world around us sees some animals as more important than others. We're taught to take good care of our pets, but to ignore what happens to animals seen as products for us to use. Many shops, businesses and advertisers don't want us to consider how meat gets to supermarket shelves, what goes into their cosmetics or where their plastic packaging ends up. They just want us to buy their products.

Whether or not we have animal companions, the decisions we make every day of our lives - what we buy, eat, use and discard - have an impact on the environment and animal life. We can choose to support cruelty and waste, or we can choose kindness and responsibility. My own aim is to live in a way that causes as little harm as possible to animals and the environment. I don't always succeed, and don't think anyone can in our society, but I do try.

If this is your wish too, you'll face some tricky decisions and some grey areas. In this book I don't aim to give correct answers - you'll need to decide for yourself, think about how your choices will affect your family and daily life, and what you can realistically manage. Even when you make your own rules, it may prove impossible to stick to them all the time. But it doesn't have to be all or nothing; small changes can make a difference.

Things can and do change. Fifty years ago it was acceptable for fashionable people to wear coats made from animal skins such as beaver, mink and fox; now many top designers have pledged never to use real fur. Vegetarians and especially vegans were once in a tiny minority - now we're mainstream. Everyone knows about the harm caused by throwaway plastics; there are climate change protests in many countries, and countless campaigns to protect animals and habitats. We've realized at last that we can't carry on treating the natural world as if it's ours to waste and destroy.

What I want to do in this book is to look at the way our daily lives affect animals, examine how the choices we make when we go shopping or on holiday can make a difference, and how we can avoid exploiting animals by choosing kindness, not cruelty. I want to look at what we buy, wear and use, and what we throw away. And I want to explore how we can do better.

If you care about animals, I hope to show how simple changes in your life can make a difference – and how you might even influence other people. Every person who chooses cruelty-free shampoo, or avoids buying plastic, or decides to cut down on meat or stop eating it altogether, can begin to influence others. Change happens slowly – sometimes much too slowly – but it does happen. And that's thanks to people who care about the world and its future. People like you and me.

ANIMAL RIGHTS

“

AS HUMANS WE’VE GIVEN OURSELVES POWER OVER EVERYTHING ELSE THAT LIVES.

Asking if animals have rights is a complex moral, philosophical and even legal question.

If we ask whether animals have the same or similar rights as humans, including the right to live free and without mistreatment, the answer from the current world would have to be a resounding *No* - they don't. As humans we've given ourselves power over everything else that lives.

This belief in human superiority is challenged by many in the environmental movement. But, as things are in law and custom, humans have assumed that we have the right to breed animals and own them, to kill and eat them, wear their skins and use them for medical research.

Perhaps the question we need to ask is: *should* animals have rights?

Peter Singer, a philosopher who concerns himself with the ethics of how we treat animals, uses the term *speciesism* in his book *Animal Liberation*. Just as racism, sexism and ageism assume that some people should have more rights than others, speciesism assumes that one species (humans) is more important than all others on Earth.

Singer says that even the word 'animal' is biased, as we nearly always take this to mean 'non-human animal' - although of course we're animals, too. The phrase 'behaving like animals' is used insultingly, meaning that someone's behaviour (or more likely it will be used about a group of people) is rough, brutal, deliberately cruel or sadistic. It shows that we think we're above the rest of the animal world, purely because we belong to the species *Homo sapiens*. And in so many ways we use this to deny other animals any rights at all.

It's argued that animals can't have rights because they can't understand the concept of morality, or because they don't have the intelligence of humans. But we accept that a newborn human baby has rights without being expected to have any idea of morals or responsibility. And the intelligence argument hugely underestimates the cleverness of, for instance, dogs, dolphins, pigs, primates and some birds.

Laws regarding animals' rights are contradictory - some are protected, while others can be killed indiscriminately. Penalties for deliberate cruelty are often too light, and bans against keeping animals aren't imposed often enough on those found guilty of abuse.

It took until 2009 for the European Union to declare that farmed animals are 'sentient beings' - that they have feelings and can suffer. Until then, farm animals were classed as agricultural products, like cabbages or sacks of flour.

Welfare organizations around the world work hard in their efforts to ensure that animals - on farms, in zoos, in the wild, kept as pets - are treated with care and consideration. Yet this is always in conflict with the desire to exploit animals for profit and the need to provide food as cheaply as possible to a world hungry for meat and dairy products. In too many countries, animals have little or no protection in law.

I honestly can't see a future in which animals are given the same rights as humans - and if they were, it would lead to further problems. What do we mean by animals? *Which animals?* If they were all given rights, would the right to live without persecution extend to fleas, locusts, malaria-spreading mosquitoes and the worms that live in the intestines of cats and dogs if not controlled? They're all animals. A line would have to be drawn somewhere – but where and by whom?

Protecting garden insect life is one thing, but standing up for locust rights would be quite another. While I was writing this book, vast swarms of locusts were devastating crops in Ethiopia, Somalia and Kenya. Farmers and smallholders could only watch in horror as the crops they'd been tending were devoured in less than a day. This was a huge crisis in terms of feeding the people of that continent. Maybe a very strict vegan would argue that the locusts shouldn't be killed, but in the face of such terrible, widespread human and animal suffering that feels like a harsh line to take.

In general, though, surely humans have the duty to treat animals as sentient beings – creatures that can feel emotions such as joy and exuberance, grief and terror and pain. Even if we consider ourselves superior to other animals, we should accept the responsibility of looking after the natural world and not see it as ours to exploit and destroy, as if we are a swarm of locusts devouring everything in our path.

And surely animals have the right to be treated with consideration. Even if they can't have the same rights as humans, it doesn't follow that they have no rights at all, or that they exist solely for us to use. My view is that we should do all we can to respect their existence, preserve their habitats and avoid causing suffering. That means thinking about our daily lives and habits, whether or not we're in direct contact with animals.

Peter Singer says in *Animal Liberation* that he doesn't think of himself as an animal lover – in that he doesn't own pets and isn't even particularly interested in animals as a naturalist or zoologist would be. He's concerned for animal rights just as he's concerned for the rights of human beings he doesn't know personally. It's a question of morality and of living in the best way we can, to cause the least harm.

That's what this book is about. Living kindly isn't only for those who love kittens or koalas, ponies or polar bears, or for the many of us who have pets or enjoy wildlife documentaries.

It's a choice for everyone who cares about fairness and the avoidance of cruelty.

“

IF WE CONSIDER OURSELVES SUPERIOR TO OTHER ANIMALS, WE SHOULD ACCEPT THE RESPONSIBILITY OF LOOKING AFTER THE NATURAL WORLD.

SPEND KINDLY

TOGETHER WE CAN MAKE A DIFFERENCE.

”

Whenever you buy something to eat, wear or use, you have the power to choose one thing over another. Many of these choices affect animals and the natural world, sometimes in ways that aren't obvious.

Choosing one brand of shampoo rather than another is a small gesture, so tiny that you might wonder if it makes the slightest difference. But together we can make a difference and change attitudes. A good example of this is the spread of veganism over the last few years. Until recently, vegans were in a tiny minority; now, vegan options can be found in supermarkets and food outlets almost everywhere.

More people are now aware that we can choose cruelty-free toiletries and make-up. Almost all of us use toiletries of some sort - shampoo, shower gel, deodorant, toothpaste or sunscreen. Traditionally, products like these have been widely tested on animals - shampoo dripped into rabbits' eyes, laboratory rats force-fed products like face powder until a 'lethal dose' could be gauged. More often than not, people don't give this a thought when shopping for eyeliner or conditioner. But what if we all did, and refused to buy products that involved animal suffering? Manufacturers would soon change their policies - and many of them are changing already, including high street brands.

Fashion, adverts, packaging and what friends use can influence us a lot - it's easy to not consider how animals are involved. But if you want to live cruelty-free, thinking about animals will become a habit. I'm not bothered by fashion and trends and for me the number one priority when buying shampoo is: is it cruelty-free? I won't buy it unless it's a brand I can trust. My next priorities are to avoid palm oil (because rainforests are often destroyed so that palm oil can be grown on the cleared land) and to choose plastic-free by going to our local zero-waste shop where

I can refill my bottle and avoid buying a new one. There's more about cutting down on plastic and what this has to do with animals in the chapters that follow.

SO HOW WILL YOU KNOW WHICH PRODUCTS ARE TRULY CRUELTY-FREE?

If the packaging makes no reference to animal testing then it's very possible that some level of cruelty was involved in its development, but you can use one of the websites I suggest below to check. You may see wording such as *Against Animal Testing* or *Cruelty-Free* on products. This is better than nothing, but in fact any company can make such a claim without meeting any particular standard. The claim may also refer only to the finished product - meaning that the suppliers of ingredients in the product may have carried out tests on animals.

I always look for the Leaping Bunny or another symbol I know I can trust. As well as the organizations I mention here, there may be others that operate in your particular area or country, so these are worth searching out online. Resources are always changing, so do keep an eye out for new ones - but do your own research too!

The Leaping Bunny
This symbol on a packet or bottle guarantees that the product has not been tested on animals at any stage of production. The Leaping Bunny logo is issued by the Coalition for Consumer Information on Cosmetics (CCIC) in the US and Canada. CCIC certifies companies to be cruelty-free using its *Corporate Standard of Compassion for Animals.* CCIC's aim is to make shopping for animal-friendly products easy and trustworthy. This is an internationally-recognized symbol, guaranteeing consumers that no new

animal tests were used in the development of any product displaying it.

Find out more at www.leapingbunny.org

Cruelty-Free Kitty

Cruelty-Free Kitty has a useful online database of more than 700 brands. To be classed as cruelty-free, brands must guarantee that:

- They don't test finished products on animals.
- Their suppliers don't test raw materials or ingredients on animals.
- No one else carries out tests on their behalf.

Cruelty-Free Kitty also includes a list of companies that, at the time of writing, did test on animals – these are mainly giant corporations that own many brands. Search the databases at www.crueltyfreekitty.com.

Look out for The Leaping Bunny logo shown here.

There's no doubt it can be confusing, and it can become very complicated if you start looking at the precise meanings of the various symbols and how they're awarded. A brand may pass one test but fail another – for instance, a face cream might be cruelty-free, but the brand is owned by a parent company that makes things like household cleaners which are tested on animals. Another product may be free from animal testing but not vegan, which could be important to you.

The key thing, though, is to look and be aware – rather than buying a new shampoo or blusher because you've seen an advert, you like look of the packaging, or someone on social media was using it.

If the packaging doesn't give the answers you need, ask questions! I often ask staff in shops and department stores what their company policy is on animal testing. Many times I've been surprised that they didn't know, but even so it's worth asking, to show that animal testing is a customer concern – and it's something they really should know about the cosmetics they're selling!

As consumers – of food, of toiletries, of clothes and shoes – we have this small power. You might not be the person who shops for food, shampoo and other products in your home, but you can still have a conversation about what you want and don't want.

Choosing cruelty-free toothpaste, shampoo and other toiletries is one small change you can easily make. If you're buying soap, shower gel or other toiletries as a gift for someone else, choosing cruelty-free items can help spread the word, too – especially if you tell them why you've chosen that particular brand.

CHOOSING CRUELTY-FREE TOILETRIES IS ONE SMALL CHANGE YOU CAN EASILY MAKE.

WHAT'S ON YOUR PLATE?

“

IN THEIR LIFETIME, THE AVERAGE MEAT-EATER WILL CONSUME OVER 10,000 ANIMALS.

Most of us are brought up to accept that meat and other products from animals are part of our meals. In most cases we start eating meat before we even understand that it's the flesh of animals.

While I was writing this, newborn lambs were out in the fields, playing in groups, exploring, resting in the sunshine. It's always a lovely sight, signalling the beginning of spring in the UK; but many families taking their children to see those animals will happily eat lamb for lunch or dinner.

How do we learn to do that - to pretend not to know that those lovable young animals will soon be slaughtered for our plates? And *why?*

I was brought up to eat meat, but gradually felt that I couldn't call myself an animal lover while sitting down to eat a Sunday roast lunch, or bacon at breakfast. I became vegetarian during my student years, and haven't knowingly eaten meat since; I'm now vegan (though not as strict as some - more about that later).

There's no question that I'll ever eat meat again, but I don't expect to persuade everyone I know to become vegetarian or vegan. Members of my family still eat meat, and although I wish they didn't, I don't choose to have a blazing argument every time we sit down to a meal. My hope is that more and more people will take notice of the compelling reasons to reduce their meat-eating, at the very least.

Like many people, my first reason for becoming vegetarian was in protest against animal cruelty. More and more, though, we're hearing of the environmental impact of rearing animals for meat, and that a plant-based diet is far more sustainable. Also, there's medical evidence that meat-eating, particularly red meat such as beef and lamb, can contribute to illnesses such as cancer and heart disease. But I'll start with the cruelty aspect.

We'd be horrified if a neighbour killed their dog to eat it – yet in supermarkets we see meat from cows, pigs and sheep cut up into pieces and labelled *cutlets, chops, sirloin, breast, thighs*. What's the difference? Advertisements and packaging train us, as we grow up, to ignore the fact that these chunks of meat come from animals as capable of feeling as our cats and dogs. I think that's why (in English at least) we talk of *pork, ham* and *beef*, rather than saying *I'm having pig for lunch.*

If you eat meat on most days of the week, how many animals will you eat during your whole lifetime?

The Veganuary organization worked out the numbers in the graphic, opposite, using information from the British government's Department for Environment, Farming and Rural Affairs (DEFRA) .

That's 10,252 animals eaten by *one person*. Imagine those animals lined up in front of you! And other countries eat even more than the UK – the World Atlas website lists Australia as the country that consumes most meat per person, with the average Australian eating 93 kilograms (205 pounds) of meat each year. The USA comes a close second, at 91 kilograms (200 pounds). For comparison, most of the countries with the lowest meat consumption are in Africa, with India and Bangladesh consuming even less. Of course religious, cultural and traditional factors are involved here, as well as farming practices, land use and availability of meat.

THE AVERAGE MEAT-EATING PERSON IN THE UK WILL EAT, DURING THEIR LIFETIME...

3
COWS

5,668
FISH

19
SHEEP

11
PIGS

19
DUCKS

3,275
SHELLFISH

21
TURKEYS

1,190
CHICKENS

If you do eat meat, which of these attitudes is closest to yours?

I like the taste of meat, so I don't mind if animals have to be killed for me to eat them.

I don't think about it much. What's on my plate is something called 'meat', but I don't want to spend time imagining how it got there.

I've decided to carry on eating meat, but less of it than I once did, and more selectively. I want to be sure that the animals on my plate were at least reared in the best possible conditions, not on factory farms.

Animals are bred to be killed, so it doesn't matter how they're treated. (I really hope you *don't* think that – but if you do, please read on anyway and I'll try to change your mind.)

I don't like cruelty but I believe that you need meat to be healthy.

I feel uneasy eating meat, and am wondering whether to cut down or stop.

BUT WHY EAT MEAT?

More and more, I think we should look at it from this perspective. Not: why *don't* you eat meat? but why *do* you eat it? Vegetarians and vegans are often asked why they *don't* eat meat, but meat-eaters are rarely asked why they *do*.

I asked several friends and acquaintances whether they feel that their meat-eating needs to be justified. The responses were interesting. Some answered jokily, some were offended by my question and a few refused to answer.

“

VEGETARIANS AND VEGANS ARE OFTEN ASKED WHY THEY DON’T EAT MEAT, BUT MEAT-EATERS ARE RARELY ASKED WHY THEY DO.

Those who did give thoughtful and detailed answers admitted that there are difficulties: they like to think of themselves as kind, responsible people, and several would call themselves nature-lovers, yet they're prepared to eat animals that have been reared and killed for that purpose. Most people don't seem to give it much thought.

So, how about you? If you do eat meat, could you take a few moments to consider why? Is it perhaps one or more of the following reasons?

- You've always eaten meat.
- Someone else shops and cooks for you, and meat is what they buy and prepare.
- You like meat, and it's part of your favourite meals.
- You wouldn't know what to eat if you didn't have meat.
- You think meat is good for you - an essential part of your diet.
- You think giving up meat would be too difficult.

And, depending on your answers above:

- Do you actually *want* to eat meat, or is it a habit you'd like to change?

Imagine that you've never eaten meat in your life (and very possibly some readers haven't). When you sit down for lunch, someone says, "Want to try this for a change? It's the flesh of an animal that's been killed specially." Or what if the tasty lunch dish you were offered was casseroled dog? Dogs are widely eaten in some countries. Quite likely you'd react with horror and repugnance. Most people don't respond like this when faced with more 'traditional' meat because habit, tradition and custom make it seem quite normal to eat parts of dead animals.

Will that change? I like to think that in a generation or two - so by the middle of this century - people will look back with disbelief and horror at our widespread habit of eating meat.

And feeding a growing population with meat has led to...

FACTORY FARMING

We're brought up to have an idyllic idea of cows in fields with their calves, happy pigs and goats, mother hens with their fluffy yellow chicks, ducks dabbling on a pond. Perhaps as a small child you had picture books that showed farmyards in this way. Maybe the word 'farm' makes you imagine this kind of cheery scene.

Meat producers like to show photographs of cattle grazing in flowery meadows - who could object to that? But although some dairy cows and beef cattle might live like this, they're in a minority.

There are some farmers who undoubtedly love their animals and give them the best care they can. On their farms, animals are known individually and treated with kindness (until they're slaughtered, that is). This is the traditional way of farming - small scale, the animals killed as close to home as possible, and the meat supplying local butchers. Eating meat can never be completely humane, but meat reared in this way is the nearest thing.

But modern intensive farming, or 'factory farming', is very different. As the name suggests, animals are treated as items on a production line. Their lives are short and they're often bred selectively to give the most meat, the most milk, the most weight gain in a short time... Profit is the number one priority. Nearly 70 billion animals worldwide are reared for food each year, with two out of three in intensive systems like this.

To maximize profit, animals are crammed together in unnatural numbers. Because pigs can injure each other when kept like this, they may have their teeth clipped and tails cut off, without pain relief. Chickens, kept in numbers of up to 50,000 in sheds, routinely have their beak-tips sliced off to stop them from pecking each other. Those bred for meat are fed on a diet that makes them grow fast - so fast that, sometimes, their legs can't even support them. Chickens kept for egg laying are hybrids, bred to lay as many eggs as possible throughout the year. As soon as they stop laying they're killed, and most are then processed as pet food.

Animals crowded in with each other like this are also given antibiotics to prevent diseases which spread quickly through intensive farms. With antibiotics used so widely like this, microbes build up resistance to the drugs which means that the antibiotics for treating humans could become less effective. Over 70% of all antibiotics used worldwide go to farmed animals. Can we ignore the potentially devastating effects to human health on top of the horrific treatment of animals?

THE FIVE FREEDOMS

The Five Freedoms, listed in 1979 by the Farm Animal Welfare Council, were drawn up by vets and have been adopted by the Royal Society for the Prevention of Cruelty to Animals (RSPCA), the American Society for the Prevention of Cruelty to Animals and the European Union. They state that every farm animal should have:

1.

Freedom from hunger or thirst by ready access to fresh water and a diet to maintain full health and vigour.

2.

Freedom from discomfort by providing an appropriate environment including shelter and a comfortable resting area.

3.

Freedom from pain, injury or disease by prevention or rapid diagnosis and treatment.

4.

Freedom to express (most) normal behaviour by providing sufficient space, proper facilities and company of the animal's own kind.

5.

Freedom from fear and distress by ensuring conditions and treatment which avoid mental suffering.

Does modern intensive farming give animals these freedoms? Clearly not, when we look at chickens packed into barns in huge numbers, pigs kept indoors without ever seeing daylight or rooting in mud, and ducks that never swim in water for their six short weeks of life. Freedom to express normal behaviour? Freedom from discomfort, pain, distress? What do you think?

Factory farming is invisible to most of us. We don't see it if we drive or walk in the countryside – we can't see through the walls of those huge barns and broiler houses. So it's easy not to realize that most of our meat comes from systems like this. And for the millions of people who live in cities, what happens on 'farms' is even more distant.

Who's to blame for factory farming? It's not just farmers – it's governments, supermarkets and everyone who buys food. That means us. Unless we choose otherwise, we're part of that supply chain. If we want to buy lots of cheap meat, farmers and supermarkets will supply it. If we buy only higher welfare meat, they'll provide that. If we don't buy meat at all, demand will drop and fewer animals will be bred.

BUT HOW DO WE KNOW WHAT WE'RE BUYING? WHAT DO FOOD LABELS TELL US?

Terms like 'farm fresh', 'country fresh' and 'natural' mean next to nothing. A 'farm', after all, could describe a barn crammed with 10,000 chickens. Some food packaging wants us to think that cheery, smiling pigs can't wait to be turned into sausages, that fish swim willingly into ocean pies or that plump turkeys look forward to Christmas and Thanksgiving. Sometimes turkeys or pigs are pictured with napkins round their necks, as if they too will be joining the joyful feast.

Compassion in World Farming and other animal welfare organizations want food labelling to be stricter and clearer. It's misleading to show a picture of a lush meadow on a packet of beef mince if the animal has never seen a blade of grass, let alone a field to graze in. Pictures of pigs with curly tails hide the fact that many pigs in intensive systems have been mutilated by having their tails cut off.

Will meat producers and supermarkets agree to this clarity of labelling, though? Probably not. Like many cosmetics manufacturers, they'd prefer you not to think about what happens before the product reaches the shelves.

Eggs in many countries now have to be labelled with the method of production, so 'free range' or 'organic' eggs are best. With pork, 'outdoor-reared' means that the pigs have spent at least some of their lives in the open air, on grass or mud. 'Pasture-fed' is best for grazing animals, but that's a luxury few shoppers can afford or are willing to pay for.

If you're concerned about where meat comes from and what sort of lives the animals on your plate have had, it's better to eat less of it and to pay more for high welfare meat.

Some cruel practices have already been banned, thanks to the efforts of animal welfare organizations and the many people who support them. Egg-laying chickens once spent their lives tightly-packed into battery cages so small that they couldn't stretch their wings: the space for each chicken to spend its whole life would be only slightly more than this book opened out flat. Battery cages are now illegal in the UK and many other countries.

Until recently, sows could be kept in crates too small to allow them to turn round, with slatted floors for their dung and urine to fall through - so they had nothing to do but eat, no stimulation, no company of other pigs, no exercise. This practice is now banned in the UK, Europe and Canada, and in some USA states.

There is still much to be done worldwide to raise awareness, with many countries having little or no protection for farmed animals.

The COVID-19 pandemic has led to a new awareness of the risks posed to humans by factory farming. Many infectious diseases are first found in animals - such as Ebola, SARS, swine flu, bird flu and the Zika virus - and can lead to pandemics with huge loss of human and animal life. Such diseases can also be spread very quickly at 'wet markets' where keeping and slaughtering animals not only raises ethical issues but poses great risks to human health.

CAN SLAUGHTER BE 'HUMANE'?

If you have a cat, dog or other pet, the time may come when you must make the tough decision to have it 'put to sleep'. It's the responsible thing to do to end an animal's suffering when no more can be done to help it.

Having owned many cats over the years, I've taken that sad trip to the vet several times. It's always a sorrowful occasion, and taking that decision, on a vet's advice, never gets easier – but I consider it my duty not to let my pet suffer unnecessarily towards the end of its life.

Would you think of sending your cat or dog to a slaughterhouse when the time came to end its life? Would you hand it over to be processed in a line of other cats and dogs – terrified because they can smell blood and sense fear – to be stunned (if it was lucky) before being hung up by its back legs to have its throat slit? Of course not. And the animals destined for slaughter aren't there because they're old or ill and have reached the end of their natural life-span, like your dog or my cat. Food animals like pigs, lambs and chickens are often killed at just a few months old.

Most meat-eaters don't want to think about what happens in slaughterhouses – and they don't have to, because meat arrives in supermarkets neatly packaged and labelled. Even in well-run slaughterhouses there is terror and suffering for animals. There may be occasions when stunning doesn't work, so the animal is fully conscious at the moment of death.

Compassion in World Farming has a number of concerns about slaughter, which include:

- Pigs being gassed with carbon dioxide as a means of stunning them before slaughter. CO2 gas produces a burning and then drowning sensation, causing severe suffering before the animals become unconscious.
- Around 1 billion chickens a year in Europe alone are ineffectively stunned before slaughter. They experience a painful electric shock that fails to properly stun them, then the pain and terror of being slaughtered while conscious.

“

THE PLANET CAN’T SUPPORT EATING MEAT AS THE NORM.

”

- Every year over 2 million animals are exported live from Europe, and millions more from Australia. The campaign against live exports is a key one for Compassion in World Farming.
- Of the billion or so fish killed in Europe each year, most are slaughtered in ways that are inhumane. Many are simply left to suffocate. There's also huge loss of life through 'bycatch' – sharks, dolphins, turtles and unwanted fish that are caught up in netting and simply thrown back into the sea, dead or dying. Among these unfortunate creatures are species in serious decline, their lives wasted by indiscriminate commercial fishing.

As if this wasn't bad enough, it's not hard to imagine that workers, expected to 'process' thousands of animals every day, break and ignore rules for the sake of speed. Even worse, some (a tiny minority, I hope) treat animals with contempt and indulge in what amounts to torture. There are well-documented accounts of abuse and deliberate cruelty in abattoirs, far too distressing for me to describe here. Some of these accounts came from slaughterhouse workers who were sickened by what they had to do and what they saw others do.

In 2018 it became compulsory to install CCTV in abattoirs in England, which should at least act as a deterrent, but this needs to be extended, ideally worldwide.

Another compelling reason to look critically at the vast scale of meat production is that...

MEAT COSTS THE EARTH

In the twenty-first century, with the urgent threats of climate change, rising sea levels and wildlife and habitats disappearing, it's clear that the planet can't support eating meat as the norm.

Of course there are countries such as India where meat-eating isn't the norm, but in most affluent western countries, meat is seen as a staple, the centrepiece of most meals. It's a deeply-ingrained habit, and a costly one for the earth.

Partly in order to satisfy our desire for meat, rainforests in South America are being cut down at an alarming rate. Not only does this deforestation reduce the potentialof forests to slow down climate change by capturing greenhouse gases, but the uses of the cleared ground are environmental problems too.

A lot of the ground is used for cattle pastures. The carbon emissions of cows and sheep are particularly high because - to put it bluntly - they burp and fart methane, which is a greenhouse gas 23 times more potent than CO2. That's why the carbon footprint of beef and lamb is higher than pork and chicken.

Much of the cleared ground is used to grow soya, too. Most of the world's soya crop is fed to animals, which are then eaten by humans. That soya is a huge food source that would go a lot farther if we ate it directly. In fact, meat has a huge environmental cost to produce in terms of water as well as soya and other crops. If you eat beef for example, you're consuming not only the meat itself but, indirectly, a proportion of the food and water the cow ate and drank between birth and slaughter.

WHY ENVIRONMENTALISTS DON'T LIKE MEAT

IT TAKES

15,000

LITRES

OF WATER TO PRODUCE 1KG OF BEEF

BUT ONLY

1,250

LITRES

OF WATER FOR 1KG OF MAIZE OR WHEAT

IT TAKES UP TO

20KG

OF GRAIN

TO PRODUCE

1KG

OF BEEF

=

IT TAKES

2KG OF GRAIN

TO PRODUCE

IT TAKES BETWEEN

4-6KG OF GRAIN

TO PRODUCE

1KG

OF LAMB

1KG OF BEEF CAN ACCOUNT FOR UP TO = 1,000KG OF GREENHOUSE GAS

OF GREENHOUSE GAS EMISSIONS WORLDWIDE ARE PRODUCED BY FARM ANIMALS

Data from 'Meat and Greens', The Economist, December 31st 2013

There's another environmental problem when it comes to factory farming - the waste products of the meat industry. Units containing thousands of pigs, chickens or indoor-raised cattle produce vast quantities of dung and urine which, combined, are called slurry. That has to go somewhere, and where it goes is into huge lagoons of waste. Slurry produces ammonia, which adds to air pollution. Despite government rules about safe storage, there are slurry spills into streams and rivers. There, it's a serious pollutant, a danger to aquatic life and vegetation, as well as possibly to humans.

CAN YOU MAKE RESPONSIBLE FOOD CHOICES AND 'EAT FOR THE PLANET'?

This means avoiding intensively-farmed meat in favour of systems that are kinder to animals and the environment. Organizations including Compassion in World Farming take the line that pasture-feeding is best. Grazing animals like cattle and sheep convert something humans can't eat, the grass, into something they can: the resulting meat.

I don't take the line held by some stricter vegans that there should be no animal farming at all. Some kinds of land benefit from being grazed by animals - for instance water-meadows, which aren't suitable for growing crops or trees, but which do support a variety of wildlife and help prevent flooding, too. Moorlands are another example - high ground where crops couldn't be grown or harvested, again with their own specialized wildlife. And the idea of 'rewilding' (leaving areas of land relatively undisturbed to allow plants, birds and invertebrates to establish themselves) relies on grazing animals such as cattle, sheep or deer to keep trees from dominating. It's too simple to

“FACTORY FARMING IS BAD FOR ANIMALS, BAD FOR THE PLANET, AND BAD FOR US.

state that pastures for animals should be replaced by crop-growing or forestry – our wildlife and biodiversity would be worse off if that happened everywhere.

But the vast majority of animals reared for meat *aren't* fed on grass. Billions of animals in intensive farming systems around the world never see grass. Pasture-fed beef does have a low carbon footprint, yes – but that's a luxury, and more expensive than other meat. And for me there remains the ethical issue of killing the animal, which usually happens at the same slaughterhouse as cows and pigs from lower-welfare systems.

There are various websites and apps on which you can work out your personal carbon footprint (see, for instance, www.carbonfootprint.com). They take into account things like your diet, how your home is heated (divided by the number of occupants), what transport you use, and what you buy or throw away. What we eat and any food we waste adds up to about 20% of our individual carbon footprint, and meat and dairy are the costliest parts of our diet in carbon terms. Mike Berners-Lee, a researcher and writer on carbon footprinting, says in his book *How Bad Are Bananas?* that the average person could cut their personal carbon footprint by 25% simply by *reducing* meat and dairy in their diet.

Put at its simplest, most of us have a choice. We can eat slaughterhouse products – dead animals, many of which have led unnatural lives in cramped conditions and have died in pain and terror.

Or we can choose not to. We don't need to eat animals. It's hard not to conclude that meat production, especially of factory-farmed meat, is bad for animals, bad for the planet, and bad for us.

Or, if we do choose to carry on eating meat, we can try to eat less and focus on that which is reared more kindly, for the animal and for the planet.

MEAT – CUTTING DOWN OR CUTTING OUT?

The great thing about cutting down on meat eating is that you can decide how far you want to go. There isn't a rule book; there are no penalties if you sometimes give way or eat meat by mistake.

Any move you make towards eating less or better-produced meat is a good step to take - not least because it means you're thinking about what you eat, how it arrives on your plate and how it affects the animal world and the environment. It doesn't have to be all or nothing.

You can take a small step at first if a bigger one is too much. You may feel that you enjoy the taste of meat too much to give it up altogether. Depending on your home and family situation, it may be difficult to make a complete switch to being vegetarian or vegan, but even a small change can have benefits.

If you do still eat meat, on the following pages you'll find that there are some changes you could make, from easy to quite challenging.

CHANGES YOU COULD MAKE TO YOUR DIET:

CARRY ON EATING MEAT – BUT EAT LESS OF IT, FOR EXAMPLE BY NOT HAVING MEAT EVERY DAY, AND/OR BY CHOOSING VEGETARIAN OPTIONS WHENEVER THEY'RE OFFERED.

STOP EATING SOME KINDS OF MEAT – FOR INSTANCE BEEF OR LAMB.

STOP EATING INTENSIVELY-REARED MEAT WHERE POSSIBLE.

STOP EATING MEAT ALTOGETHER.

STOP EATING FISH, TOO – 'NOTHING WITH A FACE'.

CUT OUT MILK, CHEESE AND OTHER DAIRY PRODUCE, AND EGGS.

STOP EATING, WEARING, BUYING OR USING ANY ANIMAL PRODUCTS AT ALL – E.G. HONEY, LEATHER, SILK, WOOL.

On that list, I come somewhere between the last two. Although I call myself vegan, I sometimes buy locally produced honey, because I like to support beekeepers - we need pollinators. I don't buy or eat commercially-produced honey, but stricter vegans wouldn't eat any honey at all. I won't buy anything made from leather or silk but I do have woollen knitwear.

Eating less meat

If you and your family eat meat on most days of the week, that makes you very privileged - I don't mean because of eating the meat, of course, but because in many environments meat is expensive compared to other foods. According to Public Health England, the amount of meat eaten in the UK is almost double the world average. Australia and the USA eat even more. People in many countries simply can't afford that, or don't eat so much meat for cultural or religious reasons.

For lots of people, main meals are based around meat or fish, with the occasional exception: they decide on the main fleshy item, then think of what to have with it. You can change this by thinking of the meat (if you have it at all) as an extra, or a garnish, rather than as the centrepiece.

If you're eating out or getting a takeaway, it's easy to go meat free- even if you're with meat-eating friends or family you can usually choose something veggie or vegan. Veggie burgers and pizzas are widely available, and vegetarian and vegan choices have always featured in Indian, Chinese, Vietnamese and Middle Eastern restaurants. Meat-free meals can now be found almost everywhere, as more and more people choose plant-based diets.

Do you do any of the cooking at home? If you do, there's an enormous range of vegetarian dishes to try making, or ready meals if you don't feel up to that. You can easily find recipes and advice online or in magazines and cookery books. If someone else cooks, you could ask for vegetarian items and ingredients to be put on the shopping list, such as a vegetarian pizza rather than a meaty one, or kidney beans, mushrooms and tomatoes for a plant-based pasta dish.

A first step could be to decide that on some days you won't eat meat.

The US #MeatlessMonday campaign was launched in 2003, and has spread to 40 other countries including Australia and South Africa. The campaign highlights the many advantages of cutting down on meat: "Eating less meat and more healthy plant-based foods can help reduce the incidence of chronic preventable diseases, preserve precious land and water resources, and combat climate change." It encourages people to have at least one meat-free day each week - that's possible for everyone, isn't it? - in the hope that cutting out meat for just one day out of seven will result in people choosing healthier options every day.

Similarly, #MeatFreeMonday was launched in 2009 by Sir Paul McCartney with his daughters, fashion designer Stella and photographer Mary. Taking the pledge isn't about detracting from the food you eat, but about *adding* to it - a chance to try plant-based recipes and new ingredients.

An alternative to this could be to decide not to eat meat until your evening meal - it's easy to have a vegetarian breakfast and lunch.

Meat substitutes can help you to make the change
For some people, it's hard to give up meat because they love the flavour and texture and can't find anything else as tasty. But if your motivation is strong, you'll get over that. I used to like meat as a child, and can just about remember the taste of chicken and lamb - but I never, ever wish I could still eat it; the idea is revolting to me.

If you want them, you can find plant-based substitutes for nearly every kind of meat. Veggie burgers, vegetarian sausages and mince have been around for years, but recently more products such as vegan 'ham', 'chicken nuggets' and even 'fish' have come on to the market.

I don't like eating food that pretends to be meat. More than once I've had to check packaging to be quite sure that I'm not eating meat - some burgers and the like resemble meat so closely that even regular meat-eaters might not be able to tell the difference. I don't want that - having given up meat years ago, why would I want to pretend to eat it? But these products can be helpful for some new vegetarians who find it hard to cut meat out of their diet.

As you get used to it, though, you'll probably stop missing meat or feeling that something is lacking from your meals; instead you will find new favourites to enjoy. Meat substitutes can either replace some of the meat in your diet or help you to make the change to cutting out meat altogether.

“

THE ONLY RULES ARE THE ONES YOU MAKE YOURSELF.

Becoming vegetarian
Being vegetarian means that you don't eat animal flesh. Some people decide to eat 'nothing with a face'! Others choose a half-way position by eliminating red meat (beef and lamb) from their diet while continuing to eat chicken and fish. Then there are those who continue to eat fish but don't eat any other kind of flesh. Various terms are used such as flexitarian (for people who are cutting down) and pescatarian (people who eat fish but not meat). There are no rules about it, other than the ones you make for yourself. You could decide to become vegetarian gradually - giving up beef first (which includes burgers, mince, etc) then other kinds of meat, then fish. Most vegetarians continue to eat eggs, dairy produce and honey.

Remember that simply cutting down on the amount of meat you eat will make a big difference to the number of animals you consume during your lifetime. A small step is still a step.

"We don't need a handful of people doing it perfectly. We need millions of us doing it imperfectly." I don't know who first said this, but it's a slogan I've seen applied to zero waste and reducing our carbon footprint, and it's just as true here. It's up to you how far you go. If you've decided to take even the smallest of steps, don't let anyone undermine your efforts.

Sneaky meat!
If you want to be quite sure you're not eating animal products without knowing, you'll soon get used to looking at labels on packages and lists of ingredients, and will learn what to look out for. Soups - even vegetable-based soups - sometimes contain chicken or meat stock, though it's perfectly easy to make them without. Some salad dressings contain anchovies, so check the label.

Not all cheeses are vegetarian - many hard cheeses, like Parmesan, contain rennet from the stomach lining of calves. Pesto is traditionally made with Parmesan or Grana Padano cheese, and you wouldn't easily be able to tell from the jar if it's vegetarian or not. Vegetarian and vegan pesto are now easily available, though, so do check those labels!

Gelatine, which is found in foods like jelly, many kinds of sweets and desserts and some yoghurts, isn't vegetarian: it's a protein made by boiling skin, tendons, ligaments, or bones (basically the bits of animals that are left over from the butchery process) with water. Sounds horrible, doesn't it? Not something you want to think about when eating a marshmallow! Luckily there are now plant-based gelatines which can replace this animal protein, so you can still enjoy marshmallows, mousses and jelly if you check first.

All this checking labels may sound like a nuisance, but it won't take long to learn which products and brands are vegetarian or vegan, and what foods you like best. To be quite sure what you're buying or eating, look out for the V for Vegetarian sign, or the wording 'Suitable for vegetarians'. There are many variations on this, with some supermarket chains using their own logo, but the Vegetarian and Vegan symbols are used internationally.

If I'm eating out in a café or restaurant, I always check - unless the item is clearly labelled on the menu - whether soup contains meat stock, or desserts contain gelatine. Many caterers are now getting the message that it's pointless to use meat or chicken stock or gelatine in a dish that would otherwise appeal to vegetarians.

Desserts and cakes can be more difficult for vegans, as so many contain eggs, milk or cream. Most menus clearly label savoury items as vegetarian or vegan but don't do the same for their desserts. Fortunately things are changing there, too: many places now offer delicious vegan ice cream, sorbets and other desserts.

Is it expensive to be vegetarian or vegan?
It really needn't be. Sometimes it's assumed that you'd need to spend more on specialist or hard-to-find ingredients, but you don't need to do this to eat a healthy, balanced diet.

It's true that some vegetarian and vegan recipe books list unusual ingredients, but you don't need them – you can make a huge range of meals from easily-found, inexpensive ingredients. Vegetables, fruits, seeds, grains, beans, rice and pasta, herbs and spices and other flavourings are the staples of vegan cooking, and for those who eat dairy produce there's milk, cheese, eggs and yoghurt, too. Even if fresh vegetables aren't available where you live, you can use tinned and frozen produce.

Cookery writer Rachel Ama, author of *Rachel Ama's Vegan Eats*, emphasizes that delicious food can be made from simple ingredients found in most supermarkets. Plant-based cooking really needn't be exclusive or elitist. In fact I would argue that it can cost considerably less to feed yourself or your family when you don't buy meat.

BE HEALTHY!

The American Institute for Cancer Research says that eating mostly plant-based foods, such as whole grains, vegetables, fruits, and beans, can play a big role in preventing cancer and keeping you healthier. That's because plant-based foods tend to be higher in fibre, nutrients, and phytochemicals that may help to prevent cancer.

A vegetarian diet is not only easy to follow but can also be a healthy one, as long as you eat balanced meals. I've heard of teenagers who decided to give up meat but lived mainly on crisps, chips and baked beans. That very limited range wouldn't contain the nutrients they'd need to be healthy!

If you become vegetarian you'll probably eat more fruit, vegetables and fibre than you did before, with less of the harmful animal fat that can lead to heart disease and high blood pressure if you eat too much of it. You're not likely to worry about these problems while you're young; but a change to a plant-based diet can set up good habits for your health now and in later life.

If you have any particular allergy, health condition or a history of disordered eating, it would be a good idea to consult a doctor or dietician for advice, rather than making sudden changes to what you eat.

HEALTH PROFESSIONALS RECOMMEND THAT WE:

EAT AT LEAST FIVE PORTIONS OF A VARIETY OF FRUIT AND VEGETABLES EVERY DAY

BASE MEALS ON HIGHER FIBRE STARCHY FOODS LIKE POTATOES, BREAD, RICE OR PASTA

DRINK PLENTY OF FLUIDS (AT LEAST SIX TO EIGHT GLASSES A DAY)

CHOOSE UNSATURATED OILS AND SPREADS, AND EAT THEM IN SMALL AMOUNTS

HAVE SOME DAIRY OR FORTIFIED DAIRY ALTERNATIVES (SUCH AS SOYA DRINKS)

EAT SOME BEANS, PULSES, FISH, EGGS, MEAT AND OTHER PROTEIN

“

WE DON’T NEED TO EAT MEAT TO BE HEALTHY.

None of that advice actually requires us to eat meat to stay healthy. According to the British Nutrition Foundation, the typical UK diet includes more daily protein than we need, mostly from animal products. As I've already mentioned, eating too much meat and dairy produce, especially processed meats like ham and bacon, is linked to diseases such as cancer, heart disease and stroke - and the typical diet in the USA and Australia is even more meat-heavy than in the UK.

For most people, all the necessary proteins, vitamins and minerals can be found in a vegetarian or vegan diet, with the possible exception, for vegans, of Vitamin B12, which can be taken as a daily pill.

While I was vegetarian rather than vegan my diet was too fatty - like many vegetarians I loved cheese, and would cook with it and often go to the fridge for a quick cheesy snack. Many cheeses, as well as butter and cream, contain saturated fat, so you need to be aware of how much you eat. We all need some fat in our diets, but it wouldn't be a good idea to replace meat in your diet with huge quantities of fatty cheese.

Being vegan will require more thought to ensure that you get all the vitamins you need - but it's still possible to eat a healthy, balanced diet.

WHY GO VEGAN?

Veganism is a step further. While being vegetarian is all about food, veganism is more like a philosophy of life: the principle of not using animal products or doing anything that harms animals. Not everyone who's vegan necessarily thinks like this, though; some people are vegan for health or environmental reasons.

I admit that it took me a long time to take that extra step from being vegetarian to becoming vegan. I tried being vegan as a young adult, but found it so hard that I reverted, and started eating dairy products again. I could make meals for myself at home, but if I went out to eat, or visited a friend, I had to make myself such a nuisance (asking about ingredients, examining packaging, refusing food offered to me) that I felt it did nothing to promote my views.

Back then, it was seen as peculiar enough to be vegetarian, with veganism seen as extreme, almost a form of self-denial. All that time I felt prickles of conscience, though, feeling that it wasn't logical to abstain from meat without also cutting dairy and eggs out of my diet.

I became vegan again around four years ago. Making the change was so much easier this time and it made me wonder why I'd waited so long. Veganism is widely recognized now, with meal choices and dairy substitutes available in almost all supermarkets, food outlets, cafes and restaurants. The choice is growing all the time.

One of my favourite recipe books when I first started to cook for myself was of course vegetarian. The introduction said that it included dairy products and eggs: "which animals give freely and willingly". Hmm. But, back then, I let myself believe that.

The problem with both milk and egg production is that they can involve cruelty just as great as that involved in the meat industry.

The unavoidable issue with both mass dairy and egg production is that male chicks and calves are of no use. In intensive systems, male chicks are killed at a few days old. Like other mammals, dairy cows don't produce milk unless they have calves; but their calves are taken away from them, often immediately after birth, and are either killed or reared for veal.

Also, both dairy cows and egg-laying chickens are selectively bred to increase the amount of milk and eggs they produce, and are killed as soon as they stop being productive. As their meat won't be of the best quality, it will most likely go into pet food.

If you cut out dairy products and eggs as well as meat, you're distancing yourself from this kind of cruelty. It takes more determination, but I wish I'd made the change sooner.

Some people still think a vegan diet must be weird, almost unrecognizable to the regular meat-eater – that you'd have to cut out everything you like and go in search of peculiar substitutes. It really isn't like that: vegans can eat pizza, pasta, burgers, pies, risottos, casseroles, stir fries, roasts, bakes, cakes and ice cream... In other words, food that's not much different from everyone else's. It's just a matter of choosing plant-based ingredients.

THE FUTURE...

...Insects?

I don't want to eat insects and quite probably you don't, either. But eating them could be an effective way of supplying the world's population with protein and nutrients without rearing and slaughtering mammals or birds, or destroying habitats.

If your policy, like mine, is not to eat anything that's had to be killed, you won't make an exception for insects. I'll stay plant-based - no mealworms or ants' eggs on my plate, thank you.

But it may make environmental sense to eat insects. Yes, they would have to be farmed, and killed, so many people, like me, won't be keen. But rearing animals like pigs and cows for meat produces greenhouse gas emissions that amount to 18% of the world's total. Insect breeding releases far fewer greenhouse gases and requires less water, too.

We're illogical in what we find revolting and what we find appetizing and lots of that is down to custom and what we're used to. For example people will pay a lot to eat scallops in a restaurant, while snails are less popular. And what's the difference between eating a locust and eating a shrimp?

Many cultures already eat insects and don't see anything strange in it. Insects such as grasshoppers and beetles contain as much protein by weight as beef, more iron than spinach, as much vitamin B12 as salmon and all nine amino acids. Insects are cheap, nutritious and - apparently - can be delicious.

If we're going to find a way to feed the world without destroying the planet, insects as food may well play a part.

...Cultured meat?

Cultured meat - which is meat grown in a laboratory - may be another way forward, satisfying people's desires to eat meat but without destroying habitats or rearing animals intensively. It's already being trialled, though whether it can be produced on a big enough scale to make a difference, and cheaply enough, isn't yet known.

Would vegetarians and vegans eat it? Would they want to? Probably not, because the cultures are still started with cells from living animals and some wouldn't want to return to eating anything that tastes like meat - even if it doesn't involve cruelty and death.

Arguments will continue about how it should be labelled, and even whether it can be called 'meat'. But if it can lessen animal suffering and at the same time reduce carbon emissions, I for one am all in favour.

CRUELTY-FREE FASHION

“

IT’S FASHIONABLE TO BE CRUELTY-FREE.

A few decades ago, a fur coat was a status symbol and something only wealthy people could afford. Then, it was perfectly acceptable to go out in public wearing a mink or beaver coat - in fact the wearer would quite likely be admired for being so luxuriously dressed.

That's changed so much in recent years that you're unlikely to think of wearing a fur coat or to have the chance to buy one even if you wanted to. People for the Ethical Treatment of Animals (PETA) has campaigned against fur for more than 30 years, with top models and celebrities posing nude alongside the slogan *I'd rather go naked than wear fur.* Supporters who shed their clothes for this cause include P!nk, Dennis Rodman, Eva Mendes and Pamela Anderson.

PETA's website has a list of brands that sell vegan fashion - this includes high-street shops and also, marked with a V, companies that sell only vegan items. Some of these brands use the PETA-approved vegan label.

Many fashion designers and brands, including high street brands, have now pledged not to use fur.

You may think from this that it's rare to find a high-street chain that does sell fur. But beware - the fight isn't over! Many garments such as hooded coats have fur trims, from animals either farmed for their fur or trapped in the wild. Do check carefully, especially if you're buying from a market stall or an outlet where labelling isn't rigorous.

SAY NO TO ANGORA

Angora wool may be luxurious to wear, but much of its production comes at a high cost to the Angora rabbits who produce it. It's a soft, thick wool used in sweaters, mittens and hats, and much of it is obtained - sheared or even ripped - from rabbits kept in small cages for this purpose.

Investigations have shown the cruelty and carelessness of the shearing process, which leaves rabbits screaming in pain. They go through this torment every three months or so, and spend their lives in isolation in their cages - highly distressing for such a social animal. There are cruelty-free ways of obtaining the fur - such as waiting for when it moults naturally from the rabbit - but the vast majority of angora is not obtained in this way.

You need to check labels, as sometimes a small amount of angora is added to other wool in a sweater. I was caught out recently when I bought a lovely sweater from a charity shop. Only when I got home did I look closely at the label and found to my dismay that it contained a small amount of Angora. I couldn't in good conscience wear a garment that had been produced by such cruel means, even though I hadn't paid the manufacturer directly. I donated it to a different charity shop - at least that shop could sell it again, another charity would benefit and the garment would not be wasted.

For soft jumpers, gloves and the like there are plenty of cruelty-free alternatives such as Tencel, hemp, bamboo and organic cotton - so please don't choose Angora unless you can be absolutely certain its production has been ethical and cruelty-free.

OTHER KINDS OF WOOL

Strict vegans don't wear any kind of wool because the vegan philosophy is against using anything derived from animals. Campaigners also point to rough handling of sheep during shearing, resulting in cuts and other wounds. And at the end of their lives, sheep may be slaughtered for meat.

Cashmere is a luxury wool whose name comes from Kashmir, the disputed territory between Pakistan and India. The wool is spun from the very fine fleeces of Kashmir goats, which are also reared in China, Iran, Mongolia and Afghanistan. Although some are kept by nomadic people, the demand for cashmere is so great that they are now farmed commercially in huge flocks that can degrade the landscape. Some reports suggest that they are shorn of their fleeces in winter – when the demand for cashmere scarves and sweaters is at its height, but also the time when the sheep most need the warmth of their coats.

I'm still unsure about sheep's wool, because I recognize that sheep-grazing can be good for landscapes and habitats and for many other species. I haven't bought any new clothes made of wool for several years, but I still wear the sweaters and coat that I already own. As with many other issues involving animals, there isn't a clear right and wrong about this: it can often be a matter of weighing up one principle against another.

ALLIGATORS, CROCODILES AND SNAKES

The skins of these reptiles have traditionally been used to make shoes, bags and accessories such as belts and watch-straps. In most cases there isn't the excuse that the skins are the by-product of meat: alligators and crocodiles are usually reared purely for their skins. These fashion items sell for sky-high prices, and are carried as status symbols.

A handbag may require two or three crocodiles to be killed. Even if you were very rich, would you really want to carry a bag made from reptiles that have suffered and died so that you could look fashionable? You can find faux crocodile, alligator and snakeskin if you really like the look of it - and for a fraction of the cost of real butchered animal. However, these are still likely to be made from animal leather stamped to look like crocodile or snake, so check carefully if you want to avoid any kind of animal skin.

Which leads to:

WHAT'S ON YOUR FEET?

If you want to avoid using animal products, you'll need to think about leather. Most people use or wear leather in various ways: belts, wallets, bags, watch-straps and most commonly, shoes and boots. Leather jackets and trousers go in and out of fashion, too, and leather clothing is always popular for bikers because of its protective qualities (although there are now synthetic alternatives that offer a good level of protection).

For years after going vegetarian I continued to wear leather shoes, telling myself that it was a by-product of the dairy industry so it made sense to use it. After a while, though, when I needed a new pair of boots, the thought of paying for dead animal skin was repellent to me. I haven't bought leather shoes since, and gave the remaining pairs I had to charity shops; now I don't own anything made from leather apart from a pair of gardening boots I've had for 30 years. Maybe that's inconsistent and I should part with them, but instead I've decided to wear them until they fall apart, and then replace them with a vegan pair. (Two principles collide here: not wanting to wear leather on the one hand;

“

A HANDBAG MAY REQUIRE TWO OR THREE CROCODILES TO BE KILLED.

on the other, not throwing something away while it can still be used. My boots are too shabby and old to be sold in a charity shop, so they'd just end up in landfill.)

As with food, it's far easier than it used to be to buy vegan footwear, and many companies now label some of their footwear as vegan. This means not only that there's no animal skin involved, but also that any glue used to make the shoes or boots isn't of animal origin. There are also specialist footwear companies who produce only vegan boots and shoes.

My earlier reasoning that leather is a by-product of dairy farming was a bit naïve. Maybe it is, to some extent, but I no longer use that as an excuse: some animals are killed purely for their leather, and leather products could come from countries with low welfare standards. If I'd seen the fate of some animals used for leather I'd certainly have given it up sooner.

THE END OF FAST FASHION?

Fashion is fun - clothes are part of the way we express ourselves and our individuality. If you have to wear a school uniform for most of the week, choosing your clothes is especially important. Your style can reflect your personality, your culture, or show your appreciation of a particular type of music or outlook.

Fast fashion has become so much a part of life that we can take it for granted: that there will be new clothes in the shops every few weeks and that magazines, social media and websites will tell us that we need to keep buying. The fashion industry depends on creating new trends every season, to make us see last year's clothes as out of date, and needing to be replaced by new ones. Some influencers claim that they can't be seen in the same outfit twice - how ridiculously wasteful is that? It's so easy to buy on impulse now too, through a touch on a screen or click of a mouse.

But the fast fashion industry encourages over-consumption and waste. As the website Ethical Consumer says, an obsession with fast fashion elevates profit over people, the environment and animals.

You may wonder what this section on fashion and waste has to do with animals. The connection is that waste and over-consumption damages habitats - products or packaging can go into landfill, polluting land and water, and fabrics shed dangerous microfibres when washed. There are carbon implications too, adding to the climate emergency that threatens all life on Earth.

The good news is that there are better choices - but first, let's explore why fast fashion is so wasteful.

The supply lines of clothes are complicated: from raw materials to the making of clothes and the transport to shops, it can sometimes be difficult to track fashion's waste. We do know, however, that clothes that aren't sold within a fairly short period are often incinerated or sent to landfill; some never even reach the shops. The UK charity WRAP (The Waste and Resources Action Programme) estimates that £140 million worth of clothing goes into landfills every year, discarded by manufacturers, retailers or individuals. In a YouGov survey, 16% of people who'd thrown away a garment in the last year said that they'd done so after wearing it just once. This means that all the resources that went into the garment's production (human labour, production costs, transportation) are wasted.

Then there's the plastic. Half a million tonnes of plastic fibres are released per year from washed clothes – fabrics like polyester, nylon, acrylic and polyamide shed thousands of microfibres every time they're washed. Once plastic is in the sea, most of it stays there forever, adding to ocean pollution and the threat to marine wildlife. Look up the Marine Conservation Society and their campaign #StopOceanThreads to find out more about this.

Particles of plastic have been found in the bodies of fish, birds and mammals in some of the most remote places on Earth, from Arctic snow to the deepest parts of the ocean. New technology shows that microplastic and nanoplastic particles are in our own bodies, too.

There's also a human cost to fast fashion. Many quickly-produced garments are made by low-paid workers. Many garments sold by global brands are produced in factories where conditions are not only bad – they're dangerous. In 2013 more than 1,100 people died when a Bangladesh factory building collapsed. Those workers were producing clothes for international high street brands, sold in shops you've probably seen in your own nearest town or city.

Besides this, fashion has some serious carbon implications.

If you're asked about your carbon footprint, the first things you'd probably think of would be how much you travel in a car and how often you fly. But our carbon footprint is made up of everything we buy, do and use, and fashion is no exception.

The fashion industry is one of the biggest contributors to carbon emissions: according to a recent report from the Ellen McArthur Foundation, the textile industry emits more greenhouse gasses each year than all international flights and all shipping by sea. That's quite staggering – where does it all come from?

Fabric and fibre production is the biggest part of this, followed by transportation (shipping, driving to warehouses, sending out to individual buyers) and the resulting carbon emissions. All this for a dress or T-shirt that we might wear only a couple of times before throwing away. And what happens when we don't want our clothes anymore? Most likely they become more waste, whether they're incinerated or end up in landfill.

Problems on this scale can seem overwhelming, and the harm to our environment, human and animal life can be upsetting. But what can we do as individuals?

The simple answer to reducing your carbon footprint through what you wear is to buy less, wear every garment as many times as you can and dispose of it thoughtfully when its life is over.

It might be easier said than done to 'buy less, and make use of what you already have'. We may need to buy new clothes because we've grown or our shape has changed, or we may want to keep up with trends in fashion we see our friends wearing. But with fashion, as with everything else, what you choose to buy and not buy has an impact, and every little change helps.

And it's not all bad news. There are fashion brands that avoid animal products, choose eco-friendly fibres and fabrics, and are concerned for workers' rights in the production process, too. There are specialist brands whose first and guiding principle is ethical clothing. Some high street chains have made sustainability pledges which you can find on their websites, but it's worth doing a bit of research on these to check for greenwashing. That's when companies try to convince you they're environmentally friendly without actually making changes in how they work.

Ethically-produced clothes do tend to cost more, especially from small specialist companies, so not everyone can afford them - but we can expect prices to come down as sustainable fabrics become more widespread and easily available. Meanwhile we all have the responsibility to buy less and to make the most of what we have.

As well as fabrics made from recycled cotton and organic bamboo and hemp, some eco-friendly companies now produce clothes made from recycled plastic or polyester.

What fabrics will we see in the future? Modal is made from beech trees and Tencel from dissolved wood pulp; Piňatex uses pineapple leaves to make vegan leather and Qmonos is made from spider silk (without harming spiders). Citrus silk is being developed in Sicily, using the residues from oranges processed for juice. And I'm sure there will be more innovations like this.

When you next need to buy something, you could check first on Good on You (goodonyou.eco) - it's a useful website for news and information about eco-fashion, where you can check how your favourite brand is rated. Each brand is given a score for sustainability, how it treats people in the supply chain, and its policy on animals.

WE ALL HAVE THE RESPONSIBILITY TO BUY LESS AND MAKE THE MOST OF WHAT WE HAVE.

Here are some easy ways to reduce your carbon footprint in terms of what you buy and wear:

WEAR IT OFTEN

Buy only what you know you'll wear, and wear it often.

BE BRAND SAVVY

Look for brands that pride themselves on eco-friendliness and sustainability.

DONATE

Look through your wardrobe, drawers or shelves to see if there are clothes there that haven't come out for a while. If you no longer like them, or you've grown out of them and haven't got a younger relative who can wear them, donate them to a charity shop or a textile bank.

MEND

Mend clothes rather than throwing them out – by simply replacing buttons, mending holes, sewing up split seams, etc. If you haven't the skill to do this, either learn (it's not hard) or find someone who can do it for you.

GET CREATIVE!

Your old clothes can be re-styled by adding new trims, buttons, patchwork or embroidery, or by shortening a dress into a tunic or trousers into shorts. This is a rewarding way to individualize your garments.

DON'T WASH YOUR CLOTHES TOO OFTEN!

This may sound odd, but don't wash your clothes too often! Unless you've been out for a sweaty run or gym session, there's no need to wash clothes each time they've been worn. Wash less often and at lower temperatures, and put clothes into a special bag to trap microfibres. New clothes shed more microfibres than well-worn ones, so that's another argument for making the most of what you already have and not buying more new clothes than you need.

BUY LESS

Shops and fashion magazines are fond of showing us the latest 'must haves' - items they want us to think we can't do without. Fashion brands obviously don't want us to cut down on our buying - but we do have a choice!

SLOW FASHION

Think about slow fashion rather than fast fashion, as advocated by Friends of the Earth and Extinction Rebellion. This means paying more for better-quality items, and then wearing them more often and for longer.

SWAP

If you have friends who are a similar size to you, swap garments with them so that you can all wear something new.

USE CHARITY SHOPS

Explore charity shops! Not only is this fun and inexpensive, but you can find lovely one-offs that can't be found elsewhere, and any money you spend will benefit the charity.

WHAT A WASTE

"

WE TALK ABOUT 'THROWING THINGS AWAY', BUT THERE IS NO 'AWAY'.

Have you heard of Earth Overshoot Day? It's the point in each year by which we've used up all the food, energy and resources the Earth can sustainably provide. From that date onwards, we're using more than we have – it's like spending more money than you have in the bank.

This date each year is calculated by Global Footprint Network. For each nation, they work out the average carbon footprint of each person set against the country's biological resources – plants, livestock, crops, trees and forests to soak up carbon.

The date of Earth Overshoot Day has fallen earlier and earlier in the year since it was first calculated in 1961. The year 2020 saw a small improvement, the date falling a little later than in 2019 – that was because the COVID-19 pandemic caused a big reduction in air flights and road travel worldwide.

Earth Overshoot Day 2020 was on August 22nd. That meant there were 131 more days – more than a third of the year – in which we used more than the Earth could give. We'd need 1.6 Earths to provide enough for the way we lived in 2020. Can we do better?

#MovetheDate looks at ways in which we can all help to push that date back by living more sustainably. That section of the Earth Overshoot Day website includes several of the ideas found in this book: eating more plants and less meat, looking after nature, and avoiding waste.

In almost every aspect of our lives, we talk about 'throwing things away'. But what do we mean by that? Where do we think 'away' is?

Mainly, we mean that we get it out of our sight, put it where we don't have to think about it - usually in the bin. But everything we throw 'away' has to go somewhere - and often it's into landfill. These are enormous structures either built into or on top of the ground which are filled with domestic and commercial waste, and they're found worldwide - it's currently one of the most common ways of dealing with all the waste we create.

Once your rubbish goes off to landfill, you probably don't give it another thought - but it's still there. It can take hundreds or even thousands of years to decompose, and as the amount of waste builds up, so does the risk to animals.

Firstly, the waste itself poses a danger. I'm sure we've all seen images of birds with their feet tangled in plastic, but mistaking plastic waste and packaging for food is also a danger for animals. Seabirds have been found feeding their chicks bits of plastic from waste sites - not only does it stop the chicks absorbing nutrients from real food, but if they eat too much it can fill their stomachs and cause them to starve.

Where we keep waste is also a danger. Landfills with a lot of food waste are understandably appealing to hungry animals, and can tempt them away from seasonal migrations. The more waste we create, the more space we need for landfills - cutting down trees, clearing land and destroying habitats. The World Wide Fund for Nature (WWF) says that a loss of habitat is the largest threat to 85% of the International Union for Conservation of Nature (IUCN) Red List of threatened and endangered species

Landfills are supposed to be sited in low environmental impact areas and should have liners beneath and above the waste to prevent the dangerous effects of leaching:

that's when contaminated water from the waste escapes into waterways or the soil, where it can be poisonous for animals and humans. But not all landfills are made safe in this way, and it's not known for how long these protective measures will last. To protect the groundwater supplies and natural resources both humans and animals rely on, we'll need to reduce the amount of waste we produce.

The idea of throwaway consumerism that produces so much waste is really damaging. It gives us the idea that we can buy whatever we want and chuck it aside when we're tired of it, over and over again. The more we use, the more we make, the more greenhouse gas emissions we create, and the more harm to our climate, human and animal life.

RECYCLE AND THE OTHER Rs

Fortunately, recycling is now part of day-to-day life for lots of us. We might have recycling bins at home and we're used to seeing separate bins for plastics, paper and food waste when we're out and about. Most of us wouldn't think of dropping litter or throwing an empty drink can into a ditch - but we can still do more and waste less.

You probably know the three Rs of avoiding waste: Reduce, Reuse, Recycle. There are other Rs too: Refuse, Repurpose, Repair, Rethink. All of these can help to lower your carbon footprint and what you put into landfill. Really we should all have the aim of throwing 'away' as little as possible - it's part of living as sustainably as possible and doing all we can to protect the environment and living creatures.

Do you have different bins at home for recycling, and for general waste? The general waste goes into landfill - and that should contain only what can't be recycled or go into the green waste or composting bin. In our household it's mainly soiled cat litter, and that's in compostable bags, plus some kinds of packaging that still can't be recycled - though I try to avoid buying that.

Let's have a closer look at all those Rs.

Reduce

Many of us have too much *stuff*: things we don't use or wear, that collect in cupboards, lofts, under beds... Often we hang on to things for sentimental reasons, but much of what we hoard is just clutter: toys, games, and things we have no use for.

A good sort-out can be cheering in its way, making it easier to find our belongings and usually discovering things that *can* be used or passed on. But even more important is to reduce the things we buy in the first place - to develop the habit of buying only what we need, not what adverts and special offers are urging us to acquire.

Each November we're bombarded with special offers and temptations in the weeks leading up to Black Friday - the last Friday in November. Ethical consumer groups in several countries have retaliated by giving the date a different name: Buy Nothing Day, a protest against runaway consumerism.

Reducing what we buy and acquire means that fewer of the Earth's resources are ploughed into making new, unnecessary items, and also means that less will be thrown away, too.

Reuse

There are two simple things you can do here. Until recently it was the norm to carry water in a throwaway plastic bottle and to buy coffee in a 'disposable' cup. In the UK we use 7 million throwaway coffee cups every day, or 2.5 billion each year. And that's just one country - imagine how many are thrown away in the whole world each year! Then we learned that most of these cups aren't easily recycled after all. Although made mainly of paper, they have to be liquid-proof, so are lined with plastic polyethylene which is tightly bonded to the paper. This means that they can't go in the general recycling and can only be processed at a few specialist plants - so most go into landfill.

Many takeaway chains are developing fully recyclable coffee cups, but it's even better not to use them at all. If you have your own resuable cup - they're widely available, with a big choice of colours and designs - you won't be creating waste when you drink coffee, tea or anything else. Many cafes and stalls give you a discount for using your own cup, and others will follow.

You can also buy a water flask instead of a plastic bottle. Download the Refill app to your phone to find places to fill up your bottle free of charge when you're out and about. With your resuable cup and water flask, you won't need to buy throwaway containers for your drinks - and it's even better if you can find a cup or flask that's made from sustainable materials.

Recycle

As well as making full use of your recycling bins at home, there are many other ways to find homes for things you no longer need.

Charity shops take books, toys, ornaments and odds and ends in good condition as well as clothes. Some have specialist bookshops, too.

In many countries including the UK, the United States, Australia and New Zealand, there are Freecycle networks where unwanted items can be given away or exchanged - look for the link to your regional network. For larger things like furniture, there are various organizations that will be happy to find new homes for them. Your local authority may offer facilities for collection and recycling - their website will probably give more information about this.

Refuse

For a start, you can turn down unnecessary packaging when it's offered. You won't need a bag from a shop if you've brought your own.

Buy Nothing Day, mentioned above, is a way of refusing to take part in a mass buying binge. At Christmas there's a terrible amount of waste from people exchanging gifts they don't need - or even, in some cases, want. You could make a pact with your family and friends to exchange small home-made gifts, or only items bought from charity shops.

Repurpose

This can be fun, especially if you like making things. Rather than throwing items away, find another use for them or turn them into something else. A quick internet search will show you a range of wonderful ideas: bike parts made into a chandelier, a small suitcase used as a cat bed, shelves made from crates and even a boat made entirely out of plastic bottles. In my garden I grow lettuces in leftover sections of guttering.

There's no end to what you can repurpose if you put your mind to it - and repurposing means that you're also refusing to buy new things, reducing your overall consumption of resources, and recycling your 'waste' in new and creative ways.

Repair

We tend to be too quick to throw old things away and to simply buy new items. That, of course, is exactly what the manufacturers want us to do – and it can sometimes be cheaper to buy a new item than to get the old one mended. But look around and see if there's a community group near you that holds repair days.

With clothes, it's easy enough to do your own repairs (darning, re-stitching split seams, replacing buttons) and it can mean you don't have to part with a favourite garment.

Rethink

Summing up all of this – rethinking our whole attitude to waste, consumption, what we have, what we accept and don't accept. We need to be responsible (another R) and try our best not to throw stuff into that place we call AWAY, because there *is* no away!

Rethinking means considering the environmental impact of everything you do. That may sound like a burden, but it will soon become a habit not to be wasteful, and remember that every small action can add up. If you want the world to be a better and safer place for animals, taking care of the environment and using resources carefully should become part of your way of life.

TAKE THE ZERO WASTE CHALLENGE!

How much plastic comes into your home each week?
Make a list: food packaging, items bought from shops or online, packaging.

Is there a Zero Waste shop near you?
If so, it will sell various items like bamboo toothbrushes instead of plastic ones and bars of organic soap. You might be able to take your own containers to fill with shampoo and liquid soap.

Does everyone in the house know your local recycling rules – what goes into which bin?
Your local authority may issue a leaflet you can display on a board, to remind everyone.

How much plastic does your home throw away each week?
Look online for plastic-free companies, where you can order household items and all sorts of other things to help you avoid putting plastic into landfill. Ask whoever does the shopping in your house to have a look with you!

How much of everything does your home throw away each week – and how much can you reduce it?
Can you get your weekly 'landfill' rubbish down to one small bin-bag, or even less?

Can you, and others in your house, find ways to cut down on the plastic you buy?
For example, by buying bars of soap instead of gel dispensers, or by choosing glass jars rather than plastic when shopping.

ANIMALS
ON SHOW

“

WE NO LONGER WANT TO SEE CIRCUS ANIMALS PERFORMING UNNATURAL TRICKS.

CIRCUSES

When I was about 10, my grandmother took me, my brother and our cousin to a circus as a treat during our Christmas holidays. There's a photograph of the three of us children in which I'm holding on my wrist a sad little monkey dressed in a jacket and tasselled hat. I'm ashamed of that now, but at the time I knew no better and was thrilled. I loved animals, and was excited that horses, dogs and other animals were part of the performance.

In those days many a circus featured a 'lion tamer' act in which a supposedly brave handler would persuade lions to jump onto stands and sit or lie there at his command. Of course they'd be provoked to roar and appear savage to emphasize the tamer's fearlessness. Elephants and camels often appeared in circuses too. As a child I gave little thought to the conditions these animals endured during transit or how they were trained.

The UK's RSPCA believes that animals should not be subject to the conditions that may often come with circus life, such as frequent transportation, cramped housing, forced training and performance, loud noises and crowds of people.

Equivalent organizations in the USA, Australia and New Zealand share this view, as well as worldwide groups such as the International Fund for Animal Welfare. After years of campaigning, more than 20 countries, including the UK, India, Israel, Singapore, 33 of the United States and most of Latin America, have bans on the use of wild animals in travelling circuses; in some countries it's up to individual cities or states to decide on policies. The definition of 'wild' in these bans includes captive-bred animals and means those not normally domesticated: big cats like lions and tigers, camels, elephants, reindeer and zebras.

Tastes have changed, partly because of the excellent wildlife documentaries we can all watch on television and online, in which we see animals in their natural habitats, filmed in wonderful detail. Most of the public no longer wants or expects to see circus animals doing unnatural tricks. We realize that the animals aren't performing because they want to but because they've been goaded or bullied into doing so, and fear punishment if they don't. There are plenty of thrilling circus acts that don't involve animals.

Campaigning continues, though, for captive dolphins, orcas and beluga whales used in performances. The organization International Marine Mammal Project says that "captivity in small tanks is detrimental for these highly social, intelligent, and wide-ranging animals. They simply do not belong in captivity." Dolphins, orcas and whales belong in the sea.

ANIMAL TOURISM

Some 'zoos' - really no more than tourist attractions - offer the chance to pose with animals for photos. It's understandable that unwary visitors often jump at the chance of a close encounter with a beautiful animal, but most won't realize that suffering is almost certainly involved. The tiger cub they're posing with or cuddling may have been drugged, and may be killed or sold on when it grows too big and dangerous to be cute. Because these 'zoos' know that visitors love to see baby animals, they often breed too many, which then have to be sold on to other zoos or circuses.

It's sad that people's natural interest in animals is exploited like this. Well-meaning tourists pose with bears, tigers or monkeys, ride elephants or swim with dolphins, and enjoy posting photos of their experience on social media. But elephants giving rides may be cruelly trained and kept in restricted conditions, often shackled by the leg. Bears are usually trained through starvation and beatings. Performing monkeys often spend most of their lives in small cages, and swimming with captive dolphins can cause them distress. It's best to avoid all such activities - otherwise your interest in animals may contribute to their suffering.

Social media sites have realized this, and if you search for #tigerselfie, #elephantride or #swimmingwithdolphins you are likely to see a warning stating that animal abuse and the sale of endangered animals or their parts is not allowed on the site.

A link leads to more warnings about the temptations to take 'animal selfies', and to the websites of the WWF, TRAFFIC, World Animal Protection and National Geographic.

So maybe awareness is spreading, and more of us are realizing that a cute photo may come at a high cost to the animal.

ZOOS – PRISONS OR ARKS?

Are zoos good or bad? What do you think? There are so many arguments both for and against zoos that you may find yourself swayed first one way and then the other.

Which of these statements do you agree with?

No animal should be kept in captivity. They should be free to roam in the wild.

Seeing animals at close range and learning about them is a great way to encourage future scientists, zoologists and environmentalists.

Zoos play a valuable role in conservation.

Zoos only exist to make money from the public.

There's no need for the public to visit zoos when we have wonderful wildlife documentaries that show animals in the wild.

Zoos encourage people to engage with animals and to care about wildlife and habitats.

There are some awful zoos that neglect their animals and keep them in bleak cages without stimulation.

The best zoos provide roomy enclosures carefully designed for each species, to replicate their natural habitat as far as possible.

The first zoos or menageries, such as one at the Tower of London which lasted from around 1200 until 1835, collected exotic animals from around the world. The Tower of London Menagerie included lions, an elephant and a polar bear, displayed to an astonished public who would otherwise have seen such creatures only in book illustrations, if at all. These animals were moved to Regent's Park in the 1830s, when the Zoological Society of London was formed (ZSL). At this time there was little understanding of what these animals needed to be healthy, either physically or mentally.

When I was a child, it was a great treat to be taken to London Zoo. From those outings I remember a chimpanzees' tea party, in which an audience gathered to watch and laugh as the chimps sat at a table to share and throw food. Meanwhile, big cats such as lions and tigers paced in small, concrete-floored cages. An elephant gave rides and would accept treats such as buns from the public, but if offered a coin it would pass it to its handler. A polar bear lived in a pit with a shallow pool in it and had little to do but pace around its small enclosure.

Back then, I don't think I questioned the ways in which the animals were kept and displayed. The big cats I saw had little chance to express their natural behaviour – it's now recognized as a sign of stress, or zoochosis, when cats pace back and forth in limited confines. And we no longer want to ridicule and laugh at chimpanzees as they eat from plates and drink from teacups, as if we see them as inadequate humans – that would be demeaning both for the animals and for us.

Fortunately, attitudes to animals and how they're shown to the public in modern zoos are very different from when I was an excited child gazing at a polar bear in a pit. In good modern zoos it's unacceptable to keep animals in stark conditions merely to display them to the curious public. The best zoos do far more than that.

But first, let's look at...

The worst zoos

An internet search will bring up evidence of truly terrible, heart-rending conditions in some zoos around the world.

In the worst, animals are kept in small, filthy cages or enclosures that don't attempt to reproduce their natural habitat, and give them nowhere to hide. Animals are given unsuitable food; some even die from starvation or malnutrition. They're not adequately protected from visitors, some of whom throw food or other items through the cage bars. Nocturnal animals are exposed to bright daylight while others suffer in temperatures that are too hot or too cold for them. Highly social animals such as elephants are kept in solitary confinement, while lone animals like tigers are forced into groups. Sick animals are neglected or left to die, sometimes with corpses left in the cages.

Zoos like these are indeed prisons for animals, and can't be said to have any value in conservation or education. What could we possibly learn by seeing animals in such awful conditions? If we do see this kind of abuse, we should report it.

The best zoos

The best modern zoos have animal welfare as their top priority and are always improving the ways in which they keep animals, sharing information with other zoos around the world. They make a genuine contribution to conservation, funding international projects to support animals in the wild. The Zoological Society of London (ZSL), for instance, has projects in every continent, monitoring wild creatures and the threats they face from poaching or loss of habitat, and working to prevent the illegal trade in wild animals.

Zoos are sometimes sanctuaries for animals rescued from suffering in captivity, and those orphaned or injured in the wild who are not suitable for release. The Isle of Wight Zoo in the UK, for instance, run by The Wildheart Trust, has lions, tigers and bears rescued from circuses or other poor conditions, and exotic animals saved from the pet trade. Detroit Zoo re-homed a polar bear from a travelling circus in Puerto Rico, where temperatures were 100 degrees warmer than in its natural home in the Arctic. Taronga Zoo in Sydney, Australia, rescues and treats animals such as koalas, wallabies, bats, and platypuses caught in the devastating 2019-2020 wildfires.

If there were no zoos or sanctuaries, it wouldn't be possible to give homes to rescued animals like these, and because of their traumatic experiences they couldn't simply be released into the wild.

The best zoos have education programmes aimed at helping us to learn and care about animals - both in zoos and in the wild. Ron Kagan, Director of Detroit Zoo, says that the modern zoo should be first and foremost known for compassion. His zoo is partnered with PETA, one of the leading animal campaigning groups. He says: "If people who work in zoos and aquariums don't define themselves as people for the ethical treatment of animals - who would?"

Other zoos that regularly appear in 'World's Best Zoo' lists are Taronga Zoo in Sydney, San Diego Zoo in California, USA, Singapore Zoo, Vienna's Tiergarten Shönbrunn (the oldest zoo in the world), ZSL and Chester Zoo in the UK and Wellington Zoo in New Zealand. These kinds of lists usually

“

ZOOS ARE SOMETIMES SANCTUARIES FOR ANIMALS RESCUED FROM SUFFERING.

”

rank both animal welfare and tourist experience, so visit the zoos' own websites to see what they say about their commitment to welfare and conservation.

ENDANGERED ANIMALS

Why are so many animals endangered? And can zoos help?

You may know that the world is facing a sixth mass extinction. The last one, 66 million years ago, saw the end of the dinosaurs and is thought to have been caused by a massive comet or asteroid striking the Earth.

Now we're bringing about another devastating mass extinction all by ourselves. There's no meteoric intervention this time, just human activity and industry warming the atmosphere and destroying habitats.

Animals are endangered in the wild for a range of reasons: urban spread and road building, destruction of habitats, agricultural practice, pollution, climate change, hunting, trapping and poaching, over-fishing, competition with other species, and even deliberate extermination. We can see that most of these threats come from human actions.

Conservation Status: The Red List

If you visit a zoo, you may notice that animals' conservation status is shown on an information board: for instance 'Critically Endangered' or 'Vulnerable'. This information comes from the Red List compiled by the IUCN. They call it a 'Barometer of Life', as it contains the detailed, regularly updated information on 70,000 species of animals, plants and fungi from all over the world.

The aim is to assess 160,000 species, so there's a way to go yet.

IUCN estimates that more than 31,000 species are threatened with extinction – that's 27% of all those assessed so far, and includes 25% of mammals, 30% of sharks and rays and an even more alarming 41% of amphibians.
A terrible statistic is that over half of Europe's native trees (58%) face extinction. If these trees die out the landscape will be drastically changed, with serious implications for wildlife – so many mammals, birds and insects depend on those trees for food and shelter.

The Red List is often talked of as if it contains only endangered species, but in fact the list goes from those not in danger at all, to those already extinct, such as the dodo.

The categories are:

- Least Concern
- Near Threatened
- Vulnerable
- Endangered
- Critically Endangered
- Extinct in the Wild
- Extinct

There are two more categories too: Data Deficient (i.e. not enough is known yet) and Not Evaluated. Of course, those species not yet evaluated may also be at risk, and then there are species we haven't yet discovered, for instance in rainforests and in the deepest parts of the oceans. In 2020 a team of scientists found 20 previously unknown species (plants and animals including a tiny frog, a viper and butterflies) in a remote area of rainforest in Bolivia, South America. Although these finds are exciting, it should make us wonder how many creatures become extinct before we even know they exist – especially in rainforests, which are being destroyed at an alarming rate.

The Red List is the best source of information about numbers, geographical range and the precise nature of threats to animals, and a fascinating database to learn from. I'll say more about it in the section about encountering wildlife as the list isn't only about exotic animals, but might give information about animals native to where you live, too.

WHAT CAN ZOOS DO FOR ENDANGERED ANIMALS?

To explore this, I'm going to look at four species, why they're endangered, whether or not they can thrive in zoos, and what their future looks like.

Ring-tailed lemurs

These beautiful primates are endangered in the wild. Lemurs live wild in only one place in the world: the southern part of Madagascar, in forests and shrubland, where they feed on fruit and leaves. They've traditionally been hunted for meat, and have also been captured for the pet trade. Because of logging, deforestation and farming, they live now in scattered groups, and their numbers are decreasing.

Lemurs are highly social animals, living in groups of up to 30, so it would be cruel to keep one lemur on its own. However, groups of ring-tailed lemurs thrive in zoos and they breed well. It's possible that in future the captive population of lemurs around the world – those in responsible zoos and conservation programmes, not the damaging pet trade – could be crucial in boosting or even re-establishing the wild population.

The Lemur Conservation Network connects over 60 organizations that protect various species of lemurs, including promoting educational conservation programmes in Madagascar. Should the threats to lemurs - and the endangered ring-tailed lemurs especially - increase though, we'll need to rely on lemur populations from zoos to prevent species going extinct completely.

Polar bears

The polar bear has become a symbol for the loss and devastation climate change will bring. These bears live only in the Arctic, catching seals as they surface to breathe through holes in the ice. But as this polar ice melts and breaks up, the bears find it hard to hunt for food, and swim further and further desperately hunting for seals or walruses. You've probably seen images of pitifully thin polar bears, unable to find the food they need. In parts of Canada the bears have roamed into built-up areas, where they frighten residents and have had to be shot or captured. They are dangerous to humans and should never be approached.

The organization Bear Conservation says that there would be no polar bears in zoos in an ideal world. In their natural state polar bears roam for hundreds of miles, and cannot thrive in captivity. Bear Conservation believes that no polar bears should ever be bred in captivity - they say that polar bears, along with orcas, whales and dolphins, suffer more from sickness and psychological problems in captivity than any other animal.

So, in other words, those bears already living in zoos should not be replaced by deliberately breeding more. At the time of writing, hundreds of these animals are kept in zoos worldwide, often in climates that are completely unsuitable for them.

“

THE POLAR BEAR HAS BECOME A SYMBOL FOR THE LOSS AND DEVASTATION CLIMATE CHANGE WILL BRING.

I look back with sadness now at the polar bear I saw at London Zoo all those years ago. But if all zoos were banned from keeping polar bears, what would happen to bears orphaned in the wild, or to 'problem bears' captured in towns? There'd be nowhere for them to go, so the only answer would be to kill them - and, surely, watch the species die out in the wild. Which raises another problem...

Polar bears' icy habitat may be completely gone during our lifetimes - a recent article in *National Geographic* reported that, based on current trends, summer ice may disappear from the Arctic as early as 2035. This would mean no seals, and no seals would mean no polar bears. Is it fair to keep polar bears in zoos when we know they can't thrive there, or should we let them become extinct in the wild? It's a difficult question to answer. The best solution would of course be to slow climate change and preserve the sea ice for wild polar bears - but is that possible?

Meanwhile, one outstanding zoo keeps polar bears in what are certainly the best possible captive conditions. The award-winning Detroit Zoo has a four-acre site, Arctic Ring of Life, which includes a large pool of chilled seawater with a clear tunnel beneath it, so that visitors can see the polar bears swimming without the bears being disturbed.

Elephants

Elephants are perhaps one of the first animals that come to mind when we think about zoos, and there's no doubt that people like to see them. But the RSPCA has reported that elephants should not be kept in zoos, as they cannot thrive in captivity. Elephants are social animals living in large mixed herds of several generations - they suffer if kept alone or with only a few others. In the wild, elephants might walk around 30 kilometers (18 miles) a day, or even further if food and water is in short supply. Many elephant enclosures simply don't replicate the natural environment that elephants thrive in, which can lead to them developing physical problems as well as anxiety and depression.

Zoos that already have elephants should care for them as well as possible, in large parks rather than in small enclosures, but the RSPCA says that they should not be bred or replaced.

Partula snails

When we talk about zoo animals, we tend to think first of the big mammals - polar bears, rhinos, elephants, gorillas, lions and tigers. It's too difficult to release these animals into the wild. But zoos also do valuable work with birds, reptiles, amphibians, insects and other invertebrates, successfully releasing them to boost populations, or even to kick-start a wild population that's become extinct.

This happened recently with two species of fingernail-sized tropical snail, *Partula rosea* and *Partula varia*, returned to their home in French Polynesia after becoming extinct there 25 years earlier. Snails don't make headline news, but it's an encouraging sign for conservation work. As part of the world's largest ever reintroduction programme, zoos bred more than 10,000 of these snails for the project. It would be hard to argue that the snails suffered while they were in captivity, and breeding them helped to reintroduce the species into the wild - a success, I'd say.

BACK FROM THE BRINK

Père David's deer became extinct in the wild by 1900 in their native home, China. The species was saved by collectors such as the 11th Duke of Bedford who bought some of the remaining animals from zoos in Europe and built up a breeding herd in the grounds of Woburn Abbey in the UK. Deer from his herd were reintroduced to China in the 1980s, and their future is looking a little brighter.

Other animals which became extinct in the wild and only exist now because of captive populations are the scimitar-horned oryx in Africa, the Wyoming toad in North America, the Monterrey platyfish in Mexico and the Christmas Island blue-tailed shinning skink (a type of lizard).

If it weren't for zoological societies cataloguing and monitoring creatures like those tiny tropical snails, how would the world even know if they no longer existed? Because of this kind of work, zoos sometimes can act as arks, saving animals for the future.

SO: FOR OR AGAINST ZOOS?

Before researching this part of the book, I found it hard to make up my mind. There are powerful arguments against keeping animals in captivity, and terrible examples of cruelty and neglect; and there are certainly animals I don't like to see in zoos, such as birds of prey which should have miles of sky to fly in. I wouldn't class Dolphinariums and tourist attractions showing aquatic animals as zoos, and I certainly wouldn't ever visit them.

But I do think that zoos - the best modern zoos - play an important role in conservation. It's not enough for a zoo to claim to be doing conservation work simply by keeping animals, and the best do far more than that - funding and supporting projects all over the world in order to understand more about animals in the wild and their needs.

The British and Irish Association of Zoos and Aquariums (BIAZA) sets out

- to inspire people to conserve wildlife and habitats
- to take part in effective conservation programmes
- to provide high quality environmental education, training and research
- to set and meet the highest standards of animal welfare in zoos and aquariums and in the wild.

Beyond Britain and Ireland there's the World Association of Zoos and Aquariums (WAZA), and the European Association of Zoos and Aquaria (EAZA). They work together to develop knowledge of conservation science and the best ways to care for animals. Vets working with zoo animals become specialists in various species - their health, illnesses and how best to care for them. Care guidelines are then shared with zoos around the world, to encourage best practice. And with their expert knowledge, zoo vets develop crucial information on animal diseases that can spread to humans.

To sum up, I think it's too simple to say that animals must never be kept in zoos. I'd certainly like to see tighter regulations around the world about how animals can be kept, and which species, as well as much tougher penalties for neglect. But overall I think zoos do have a valuable part to play in ensuring the survival of threatened animals, teaching about them, and preserving their wild habitats. Focusing on charismatic big mammals can divert us from the important work zoos are doing with smaller creatures such as amphibians and reptiles. These animals often live in zoos without distress, and many have been successfully reintroduced to the wild. Without zoos and the Red List, we wouldn't have the detailed knowledge that's needed to preserve life in the wild.

“

I THINK IT’S TOO SIMPLE TO SAY ANIMALS MUST NEVER BE KEPT IN ZOOS.

ANIMALS
AT HOME

“

WE SHOULD WANT PET OWNERSHIP TO BE GOOD FOR THE ANIMALS AS WELL AS FOR PEOPLE.

It can be wonderful to get to know and love an animal, learn to communicate with it, look after it and understand its needs. Pets can be good companions, and in some cases - such as hearing dogs and guide dogs for the blind - they are a great help to their owners. And for many people living alone, the presence of a cat or dog can be invaluable. Dog owners get regular walking exercise, and there's evidence to show that stroking a cat, dog or rabbit is good for us, calming anxieties and boosting our immune systems.

But of course we should want pet ownership to be good for the animals as well as for people.

The first pet I owned was a blue budgerigar, Peter, given as a present by my grandmother when I was about six. Peter lived for years. He had a small cage but we let him out regularly to fly around the room and to have a bath, which he seemed to enjoy. I was too young to know, but it was cruel to keep just one budgie - they're highly sociable animals that live in flocks. Peter spent long periods chattering to his reflection in a mirror, apparently quite absorbed. This seemed sweet and we assumed he was happy, but now I realize that it was the only way he could give himself the illusion of having company in his cage.

Something else that was commonplace when I was a child, and until fairly recently, was giving goldfish as prizes at fairgrounds. These poor fish were displayed, each in a small plastic bag of water, and won as prizes. How many of those fish survived, and went on to be kept in anything like suitable conditions? It was cruel and completely irresponsible to hand them out to people who'd made no plans for keeping fish.

Animals should never be given as prizes, nor any animal given to someone who isn't ready for the responsibility of caring for it. That's why the UK charity, the RSPCA uses the slogan: *"A dog is for life - not just for Christmas."* Unfortunately, animal rescue organizations still take in too many pets abandoned after the Christmas holiday, animals thoughtlessly bought on an impulse or given as presents. And recently we have seen a similar situation arise as people bought pets to keep them company during the COVID-19 lockdowns, only to tire of them afterwards.

Before taking on a particular pet, it's important to ask yourself questions about what it will need and how you'll care for it.

For example, let's say you're thinking of getting a dog:

- What sort of dog are you considering? Large, small, in-between? Bear in mind some breeds of dog will have particular needs or demand more exercise than others.
- A puppy, or an adult dog? Puppies are lovely but will need extra attention - training, regular health checks, special food. And they may be destructive, chewing up shoes or furniture. And remember that a puppy may look cute, but it will soon become an adult dog.
- Where will you look for a suitable dog? Will you buy one from a breeder? Look for advertisements locally? Find a rescue dog that needs a home?
- What special equipment will you need? A bed, puppy harness, toys, feeding and drinking bowls?
- Where will the dog sleep at night?
- Are you willing to take your dog out for walks every day, including when it's cold or raining?
- Does your home or garden have space for the dog to play and exercise in between walks?
- Who'll be in charge of daily care - feeding, walking, clearing up mess?

- Do you know what the dog will need for a healthy diet, and how much food to give it each day?
- How will you train your dog to obey simple commands – sit, stay, walk at heel?
- How will you 'toilet train' your dog?
- Does everyone in the house, especially younger children, know how to behave with the dog?
- Will the dog need to be left alone for long periods, for instance in the daytime when everyone's at work or at school?
- Are there times when your dog will have to go into kennels, for instance if you go on holiday?
- How much will it all cost? Remember you'll need to take the dog for regular vet check-ups as well as worm treatments and vaccinations – it all adds up!
- Will your dog be neutered? (i.e. spayed or castrated so that it can't breed). There is plenty of advice on this from welfare organizations.
- Are there already other animals in the house, for instance a cat? How will you introduce the new arrival?
- Have you taken into account that your dog may have a long life – 15 years or more? Can you plan for its future?

This list isn't meant to be off-putting, but thinking about these questions will make sure that you're well-prepared for owning a dog, helping it to settle into your home as smoothly as possible.

Animal charities are often great sources of advice. Search their websites to find more on bringing a pet into your home, and how to care for them.

DON'T BUY FROM A PUPPY FARM!

These 'farms' breed puppies in large numbers, often in poor conditions that don't give them a good start in life. Although you might think you're rescuing a puppy by taking it, this actually encourages the seller to breed more.

Organizations in the UK including The Kennel Club, British Veterinary Association and the RSPCA have jointly formed The Puppy Contract – a website full of information about how to spot a puppy farm, how to choose a puppy and how to care for it. Puppy farming is increasingly being banned, but if you do see an advert that makes you wary, do report it to local animal welfare organizations. In the USA, the American Kennel Club gives lots of advice about how to find a puppy that's been responsibly bred, and how to avoid 'puppy scams'. There are also organizations in other countries that can provide help and guidance.

COULD YOU GIVE A HOME TO A RESCUED ANIMAL?

There are already far more potential pets than there are good homes for them! Animal organizations can often point you in the direction of sanctuaries looking for homes for particular breeds or species.

Although I like dogs, I'm more of a cat person. Over the years I've had nine cats which came to us in various ways. The first two were strays who 'adopted' us; another two were kittens rescued by the RSPCA after being dumped. It's so rewarding to give a good home to animals that have suffered neglect. Our two kittens thrived – it helped that they were together, brother and sister. But the female, Hazel, was frightened of strangers until her old age.

“

THERE ARE ALREADY FAR MORE POTENTIAL PETS THAN THERE ARE GOOD HOMES FOR THEM.

It seems that she never forgot the harsh treatment she'd had as a kitten.

If you visit a shelter or sanctuary, you'll probably fall in love with one or more of the animals in need of a home, but you need to choose carefully. You should be aware - and the staff will make sure that you are - that some of these animals will need careful treatment. Some dogs, for instance, may have had upsetting experiences with other dogs, and can't settle happily in a house with others. Cats like our Hazel may take years to get over their fear of humans; some may show aggression when approached. Not all rescue animals have been so unlucky, though - many are in shelters because their owners have died or had to part with them for some other reason, and have always been treated kindly.

You may feel that you'd find special satisfaction in re-homing an animal that's been ill-treated - if you have the patience and calmness to do so, not rushing your new pet but letting it settle with you at its own pace. You should also get expert advice from the sanctuary that is re-homing the animal, or from an animal charity, before taking home an animal that may have been ill-treated.

There will almost certainly be a home check to ensure that you can keep the animal safely and securely. Most cat rescue organizations will require you to have the cat neutered, if it isn't already, so that it won't breed and add to the already too-large cat population.

Another way to find an animal that needs re-homing is to ask your local vet. Two of our cats, Finn and Fleur, came to us that way. Their owners were emigrating and they'd asked the vets if a local home could be found.

EXOTIC PETS

What is an exotic pet?

Reptiles including snakes, lizards, turtles, terrapins and geckos

Insects and spiders

Birds such as budgerigars, parrots and macaws

Small primates such as marmosets, capuchins and squirrel monkeys (the RSPCA wants the keeping of primates as pets to be banned)

Some people enjoy keeping unusual pets, but - as with any animal - their welfare must come first. Some have very particular needs for space, temperature, humidity and food. Sadly, too many people buy creatures like snakes, geckos and lizards without first finding out exactly what the animal needs to be healthy.

Some irresponsible owners simply dump their animals when they tire of looking after them, or the poor creature dies from neglect. Releasing them into the wild is equally cruel: many exotics will die of cold or starvation, while others could endanger wild animals or spread disease.

Sadly, the RSPCA reports that even some pet shops selling exotic creatures don't know enough to give good advice to customers, and this leads to people buying animals without the information they need. Vets say that although deliberate cruelty is rare, they see a great many exotic animals that have been mistreated through ignorance.

The bearded dragon is one of the most popular lizards kept in captivity. If you're planning to keep one, you'll need a vivarium that's at least 120cm (4 feet) long x 60cm (2 feet) high x 60 cm (2 feet) wide. This lizard's native habitat is the hot scrubland of Australia, so you must provide a guarded heat lamp with a thermostat, and both a hot area of the enclosure for basking (38-42°C / 100-107°F) and a cool area for resting (22-26°C / 72-79°F) as well as a UV lamp designed specifically for reptiles so that the bearded dragon gets the right amount of light, with complete darkness at night. These temperatures, and the humidity of the vivarium, must be checked daily. The dragon will need a place to hide, branches for climbing and basking, and sand to burrow in. As for food, it needs certain kinds of salad greens, and live invertebrates such as crickets and locusts - these will need a holding pen, plus suitable food while they wait to be fed to the bearded dragon. And if your dragon becomes ill, it will need to see a specialist reptile vet - so you need to know where to find one before buying the animal.

This basic information tells us that a bearded dragon certainly isn't a pet to buy on impulse or to be taken on lightly. All these arrangements must be ready before you bring one home.

The RSPCA rescued 145 bearded dragons in the UK in 2019, so just imagine how many need to be rescued by agencies like this all over the world. The growing trend for keeping all sorts of reptiles means that far too many are bred. This leads to breeders selling them cheaply or even giving them away, which can tempt people into having one for the novelty of owning an unusual pet.

Responsible re-homers of exotic animals will give them only to people who can promise to keep them properly, in suitable accommodation. Only think of getting one if you are certain that you can provide everything it needs and that you'll be able to devote enough time and attention to keeping it healthy - not just now, but for all its life.

PETS AND FASHION

Unfortunately, there are fashions in pets just as there are in clothes. Miniature pigs, handbag-sized dogs, meerkats and terrapins have all been objects of desire, but these trends are rarely good for the animals. Anyone who buys a pet because of a trend is unlikely to have that animal's needs as top priority.

There was mass buying of terrapins when the cartoon Teenage Mutant Hero Turtles became popular in the 1980s and 90s, but sadly huge numbers of terrapins were dumped when they grew too big or owners no longer wanted them. Even with domesticated animals like dogs, popularity can cause harm - for example, pugs, bulldogs and Pekingese are selectively bred for their flattened faces, with the result that many of them have breathing difficulties.

The demand for unusual pets is growing, largely because of social media. One example of an animal harmed by a social media trend is the slow loris, a small primate from Southeast Asia with enormous eyes. Pictures and films of these furry creatures apparently raising their arms to

“

FOR EVERY ILL OR DISTRESSED ANIMAL RESCUED FROM NEGLIGENT OWNERS, THERE MAY WELL BE COUNTLESS OTHERS THAT SUFFER UNSEEN.

be tickled went viral and demand for them soared - too many people wanted a cute, Instagram-ready pet without educating themselves first.

In fact, keeping a slow loris in captivity is cruel. They're not cuddly toys but shy, sensitive animals that don't like to be handled, and they can give a painful, toxic bite when threatened. Because of this, before they're sold many have their sharp teeth removed in a risky and painful procedure with no anaesthetic. The raising of their arms is often shown on social media by owners who fondly imagine that the loris is enjoying the attention. But it isn't! The arm-raising is a sign of fear and distress; the loris has glands in its armpits that exude toxins, to ward off attackers. And they have huge eyes for a reason - they are nocturnal, so bright daylight is painful and distressing to them.

There are some creatures that should never be kept as pets, and the slow loris is just one example of animals that cannot thrive in these conditions. Others are raccoon dogs, possums, meerkats, undomesticated cats and primates such as monkeys.

Some of the most popular species in the pet trade are marmosets, squirrel monkeys and capuchins - all of which appeal to unwary buyers because they're pretty, with big eyes, and maybe because they resemble humans and can appear entertaining. They're tropical animals that need a warm climate, specialized food and plenty of exercise; in the right conditions (not as pets) they can live for 20–40 years. They are reared by their mothers and live in sociable groups, so keeping one on its own is cruel and unnatural. Some of those on sale have been taken from the wild, with mothers often killed so that their babies can be taken - so buying one can encourage a cruel trade that's endangering wild populations.

It's sad to think about, but for every ill or distressed animal rescued from negligent owners, there may well be countless others that suffer unseen.

THE ILLEGAL TRADE IN WILDLIFE

Another worry is that the demand for exotic pets may lead to wild-caught animals being trafficked.

Some trade in wild animals is legal, but wildlife crime is big business. It's a huge black market, involving millions of animals around the world. Often it's falsely claimed that animals taken from the wild have been bred in captivity, misleading buyers.

Because of the secrecy it's hard to give exact figures about this illegal trade, but according to the WWF, it's the reason why some animals are on the endangered list. For example, the African grey parrot has almost disappeared from its native Ghana because so many are captured to be sold. They're popular as pets because of their cleverness in mimicking human voices and learning words and phrases, but this has put them at risk. The parrots don't easily thrive in captivity, and an estimated 45-65% of them die while being transported. The EU banned the import of African grey parrots in 2007, but they're still traded elsewhere across the world.

The cruel ways in which some animals are transported - stuffed into crates, boxes or even plastic bottles, sent on long journeys without food or water - are too sickening for me to describe in detail here. A high death rate in transit is simply accepted as an unavoidable 'business' loss. Birds are particularly likely to die when packed and transported without proper care.

As well as animals intended as pets, this trade includes animals destined for circuses and disreputable zoos, and also the sale of animal parts such as elephant ivory, rhinoceros horn and tiger parts for 'traditional' medicine, and reptiles for their skins. Pangolins, the most trafficked mammal in the world, are on the edge of extinction in the wild - it's estimated that a million of them have been taken from the wild in the last 10 years. They're sold for their meat and because their scales are believed to have healing powers. Pangolins are an example of an animal driven to the brink of extinction by both legal and illegal trade.

The WWF is one of the organizations determined to fight this trade and to make laws and penalties tougher. CITES - the Convention on International Trade in Endangered Species - has imposed bans on the trade in some species and limits on numbers for others. But illegal smuggling goes on nevertheless, and is the biggest direct threat to many endangered species.

So, if you do decide to buy an exotic animal, bird or reptile to keep as a pet, please make sure that it's captive-bred, that you trust the seller, and that you're not unknowingly supporting the cruel trade in captured wild animals.

If you do decide to go for an exotic pet ...
here are some questions to ask yourself first:

- *Why* do you want to keep this particular creature?
- Do you fully understand what you will need to keep the animal healthy?
- Can you afford to buy and install the specialized equipment and food?
- Are you prepared to spend the time it will take to look after this animal?
- Is there a specialized exotics vet near you?

- Will you be able to afford vets' fees if your animal becomes ill?
- Do you know how big this animal will grow, and how long it will live?
- Can you find an animal through a rescue centre?
- If you buy one, how will you check that the seller is concerned about the animal's welfare, and not just out to make money?
- Can you be absolutely sure that the animal hasn't been captured from the wild or illegally traded?

Despite all these warnings and cautions, I have to say that owning a pet can be one of the best and most rewarding experiences of your life. Loved and cared for properly, your pet will bring joy, companionship and happiness to your life and the lives of those around you. Your pet will be part of your family and you will probably have fond memories of them forever – just as I remember all the cats who've lived with me.

“

OWNING A PET CAN BE ONE OF THE BEST EXPERIENCES OF YOUR LIFE.

THINGS THAT CREEP, CRAWL, WRIGGLE AND FLY

“

A LOT OF PEOPLE SEEM TO THINK THAT EVERYTHING THAT CRAWLS, SCUTTLES OR CREEPS IS DIRTY, DANGEROUS AND DISGUSTING.

I know people, including some of my friends, who fuss and shriek when they see a perfectly harmless spider and would possibly even kill one if they found it in their home - as if a house spider here in the UK poses any kind of threat! There are spiders with poisonous bites in other countries, but none here. (If you want to remove a spider from your house but don't like touching them, you can easily trap it under a glass or mug, slide a sheet of paper underneath, then carry it to safety.)

Somehow this attitude of revulsion changes when people see cobwebs draped across dewy grass in autumn, or frosted webs in winter - then they're more likely to admire the beauty and delicacy of these spider creations.

Sadly, a lot of people seem to think that everything that crawls, scuttles or creeps is dirty, dangerous and disgusting. Worms? Yuck! Beetles? Urgh! Caterpillars? Ewww!

But get to know them and you'll soon find them as fascinating as any other wildlife. People may be squeamish about worms, but where would we be without them? If you've got worms in your garden, that's good - it means the soil is healthy. No worms would mean the garden's in trouble. The humble earthworm is what's known as a *keystone species* - it's essential to the balance of its environment, in this case by recycling leaves and other organic matter and improving soil nutrients, enabling plants to grow. And we all depend on plants.

As for insects - there are a lot of them about! According to National Geographic, there are about 1.4 billion insects for every human on Earth, and all of them play a crucial role in the ecosystem.

Where would we be without bees, for example? The Earthwatch Institute says that bees are the most important living thing on the planet. That's because so many plants depend on bees to pollinate their flowers, and that includes a vast number of plants whose fruits and seeds we eat.

Another really fascinating insect is one you've certainly seen, and can find anywhere. I'm thinking of the common garden ant. Have you ever taken a close look at ants and how they behave?

ANTS – A WONDER OF NATURE

Once, on holiday in the Italian countryside, I noticed a line of ants moving purposefully down the side of a tree-lined lane, all going in the same direction. I was curious – what were they doing, and where were they going? This line of ants stretched for more than half a mile before disappearing underground. Apparently they were moving home, many of them carrying eggs or pieces of leaf, all moving at a brisk purposeful pace.

Presumably something was threatening their colony. But who had decided that they'd move? Who knew where these ants were heading?

Ant co-operation is astonishing. If you disturb an ants' nest in your garden, as I sometimes do accidentally while gardening, you'll see that worker ants immediately get to work, some carrying eggs to safety. If they're red ants or wood ants they may bite you if you're close enough (their bite is an injection of formic acid, enough for you to notice a small sting but not harmful). A nest of ants acts as a team, each ant knowing its role and busily doing its job.

To many people, ants are just a nuisance and a pest. But if you look at them and learn about them you'll realize that they are one of the wonders of nature, here literally beneath our feet.

My garden has a lawn that contains several ants' nests. The kind of gardener who likes a perfect lawn would probably try to get rid of them. I'm horrified to see that you can buy ant killer in supermarkets and garden centres, as if that's the automatic response to seeing ants around your home. But I'd rather have wildlife in my garden. The ants cultivate the soil and help with pollination, and because of them, green woodpeckers, which eat the ants, are regular visitors.

AMAZING FACTS

The ant is one of the strongest creatures in the world in relation to its size – it can carry 50 times its own weight.

The largest known ants' nest in the world was found in Argentina – a super-colony, around 6,000km (3,700 miles) wide!

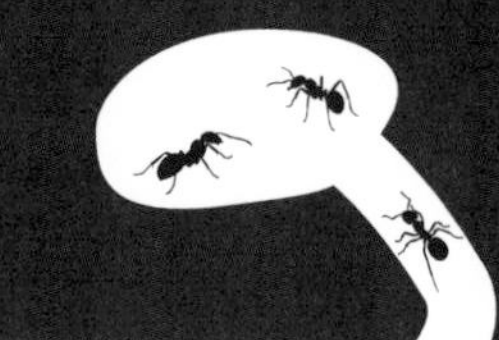

It's estimated that there are 1.5 million ants on the planet for every human being.

Ants were around at the time of the dinosaurs – 130 million years ago. That's about 128 million years before humans began to evolve.

Ants were farming long before humans were. They grow and tend fungi underground, and they keep aphids for their secretions, which provide sugar-rich food for the ants. In return the ants feed and protect the aphids.

ABOUT ANTS

They communicate through scents, touch and vibrations. Just like bees, one ant can tell other ants where food is to be found.

They live in every part of the world except for Antarctica.

Injured ants are rescued by others and taken back into the nest to recover.

By clinging together, ants can use their bodies to make bridges which other ants can cross.

Queen ants can live for up to 30 years – that's the longest life of any insect.

INSECTS ARE KEY TO ECOSYSTEMS

The insect world includes such beauties as butterflies, bees, lacewing flies, dragonflies and damselflies. There are others we're much less fond of. Wasps don't have many fans; neither do midges or mosquitoes. But they are all important to their ecosystems.

Wasps, for instance, are pollinators, and they prey on other insects and their grubs, keeping the insect population in balance.

Mosquito larvae help to keep pond water clean by eating detritus. Adults and larvae are eaten by birds, frogs, and fish, and they also act as pollinators. While some mosquitoes carry malaria, most are not harmful to us – there are more than 3,000 species of them, of which only about 200 bite humans. And there's one way in which mosquitoes have helped us: scientists studied their proboscis (elongated mouth part), to develop needles for injections that hurt as little as possible.

What *is* an ecosystem?
It's a particular environment and everything in it, both living and non-living – so that's plant and animal life, light and heat and moisture. In nature, everything plays its part: in consuming, digesting, scavenging, growing, decomposing. An ecosystem could be vast in size, like a desert, or a tiny pond in your garden. It can be on land, in the ocean, or in a stream or swamp. Each ecosystem supports its own plants, animals, invertebrates and microbes, and in turn they keep it all in balance. That's why removing one species can have a drastic effect on many others.

“

AN ECOSYSTEM COULD BE VAST LIKE A DESERT, OR A TINY POND.

KEEPING IT ALL IN BALANCE: HOW CAN SEA OTTERS REDUCE CARBON EMISSIONS?

- In the northern Pacific, sea otters were killed in huge numbers for their fur.
- Sea otters eat sea urchins. With the otters gone, the sea urchin numbers multiplied.
- Sea urchins eat kelp – underwater forests that soak up carbon from the atmosphere, shelter fish and other marine life.
- The sea urchins ate all the kelp until the seabed was as bare as an underwater desert.

BUT THEN...

Sea otters were reintroduced.

They ate the surplus sea urchins.

The kelp grew back.

Fish and other sea creatures returned to the kelp forests.

The kelp carried on soaking up carbon.

A healthy ecosystem was restored!

I like this example because it shows how things are connected: the killing of animals for fur, interfering with a habitat, and climate change. And also because - in this case - humans worked out how the damage could be undone.

Another example of balancing ecosystems is the reintroduction of wolves into Yellowstone Park, in the western United States. Because wolves were feared at the time, they were ruthlessly hunted and shot. This had a drastic effect on the elk population. With no wolves, there was nothing for the elk to be afraid of. They grazed all the vegetation from the riverbanks - grass, shrubs, seedlings - which meant there was little food, protection or shade for birds, fish and other small animals. With so little grass, soil was washed into the river and the banks crumbled away.

Then, in 1995, wolves were brought back. The elk, now fearful, sought the cover of trees rather than grazing out in the open. Plants, trees and shrubs grew again on the river bank, bringing back the birds, aquatic animals and fish that had disappeared with the wolves. Beavers returned to build dams, creating pools that drew otters and fish, and where water plants grew, attracting fish, invertebrates and amphibians. The wolves killed some of the coyotes which had preyed on rabbits, so there were now more rabbits and other rodents, attracting birds of prey. The whole ecosystem became richer and more varied.

Top predators, or apex predators, like the sea otter and the wolf in these examples, are good for ecosystems - in fact crucial to their health and balance. By killing some creatures, predators allow others to thrive. As these examples show, the removal of one key species can affect not just plants and animals, but carbon emissions, the soil and the atmosphere.

Ecosystems are always changing, even without human interference, as different species move in, trees grow taller or light levels change. But human activity such as killing animals or cutting down rainforests can have dangerous effects. Sometimes we don't know what the damage will be until it's too late.

COULD EARTHWORMS BE MORE IMPORTANT THAN GIANT PANDAS?

This idea came from Sarah Johnson, a researcher in environmental science. The idea may sound surprising, but here's the reasoning behind it.

It would be sad if the giant panda became extinct - the loss of an iconic species, an endearing animal everyone can recognize. But according to some zoologists and naturalists, the loss of the giant panda wouldn't greatly affect its ecosystem. That's because the giant panda isn't a keystone species in its environment, unlike wolves and sea otters which are top predators. Pandas are rare, solitary animals who eat only bamboo, so just about the only effect would be that less bamboo would be eaten; but because pandas are scarce and isolated, that wouldn't make much difference to the bamboo forest.

But if the humble earthworm went extinct there would be drastic results.

Charles Darwin made a close study of earthworms. In his last book, which was all about them, he wrote: "It may be doubted if there are any other animals which have played such an important part in the history of the world as these lowly organized creatures."

Earthworms are vital in making the soil healthy for plants. They recycle organic matter, fertilize and aerate the soil, helping it to drain; this reduces the risk of flooding and soil erosion. They also repair damaged soil and clean up toxins. Without healthy, fertile soil we wouldn't be able to grown and harvest the crops that feed humans and animals, so worms play a crucial role in lots of ecosystems.

Next time you see an earthworm, remember that it's one of the most important creatures on the planet - take care not to disturb it from its important work!

GO WILD

“

NATURE IS EVERYWHERE. YOU ONLY NEED THE HABIT OF LOOKING OUT FOR IT.

So far this book has focused mainly on domesticated animals and the way that humans use or interact with other creatures. However, it's important to remember that wildlife is all around us - even in cities. You can learn a lot about nature by paying attention to the wildlife in your local area - whether that's a garden, your local park or the countryside. Appreciate nature and wild animals and you'll learn more about other species, and can even contribute to the health of the planet.

WHERE TO LOOK?

If you live in a rural area, you're probably very aware of local wildlife. But even if you live in a city, there's much to discover when you're alert to it. As well as parks, rivers, canals, allotments and even nature reserves in cities, there are ways in which wild creatures live alongside humans, even below and above us. Birds roost on city roofs. Foxes are clever at finding food and dens in urban gardens. Small and larger rodents live near human habitations, often unseen. Railway cuttings and canal towpaths provide habitats for a range of wild flowers, which host a wide variety of insects.

What's in your area? An internet search will show you where to see nature in many cities and urban areas worldwide. Wetlands, rivers and reservoirs can be a particularly good place to look for wildlife, as ducks and geese out on the water are easy to see, as well as waders on shores and mud banks.

British birdwatcher and author David Lindo, known as 'The Urban Birder', specializes in showing people wildlife where they least expect it, in urban centres. Visit his website (see the end of the book) for ideas and inspiration. Perhaps there are people or wildlife groups doing this in your town or city too. Local study days, guided walks and family activities can be a great way to learn about wildlife and meet people with similar interests.

In the countryside, there are likely to be nature reserves and country parks you can visit, as well as networks of footpaths and trails through farmland, woods, moorland and coastal areas. A good map will show where you can go, as well as local walking guides.

When you look out for them, you'll also notice seasonal changes wherever you live. In the UK, the beginning of summer is marked by the arrival of swallows, and – a little later – swifts, flying in from southern Africa. Their autumn departure is followed by the arrival of 'winter thrushes', redwings and fieldfares that fly down from northern countries to spend the winter in our relatively warmer climate. Wherever you live, there will be changes like this throughout the year. If you're a keen observer you can take part in surveys which help monitor the numbers and migratory whereabouts of birds, insects and animals, providing valuable information.

If you have the chance of going on holiday, there may be good opportunities for watching wildlife – and you don't have to travel to exotic places. Are you near the coast? There may be excellent habitats such as cliffs, river estuaries or wetlands to observe wildlife in. Low tides on beaches or tidal rivers are great chances to see wading birds, and if you're able to take a boat trip then you might even see seals, dolphins or sharks depending on where you are. Further inland, there may be nature reserves, lakes and national parks to visit – try finding a guided walk or activity day to show you things and places you might miss on your own. For holidays completely focused on wildlife, there are specialist travel companies offering tours on every continent.

NATIVE ANIMALS UNDER THREAT

When we think of endangered animals, the first to come to mind will probably be big exotic creatures like polar bears and tigers. But we need to look at our own native animals, too, wherever we live.

Many regional and national Red Lists have been published around the world. In 2020 the first Red List for the UK, where I live, was issued. Alarmingly, the list of creatures under threat includes hedgehogs, water voles and red squirrels.

When I was a child, hedgehogs would regularly visit the garden; now their numbers have plummeted, and they're classed as Vulnerable. Walking by a river you'd often hear the loud plop of a water vole or maybe glimpse one. (Ratty in the classic tale *The Wind in the Willows* is actually a water vole.) It's horrifying that these much-loved creatures are being pushed out of existence, and there are examples like this in every part of the world.

What can we do?
Perhaps the most important thing is to be aware of wild animals, birds and insects near you, to learn about them and stand up for them. When you know more about these animals, you might be able to contribute to research and conservation projects, or you could help protect their habitat by helping to create a wildlife-friendly garden, for example.

Check out more ideas and organizations at the end of the book.

Nature is everywhere. You only need the habit of looking out for it.

WEBCAMS

A great way to see wide ranges of wildlife both locally and across the world is to watch webcams such as those reached through The Wildlife Trusts and other websites. These cameras aren't active all year round, but during spring and summer in the UK alone you can watch ospreys, peregrines and barn owls rearing their chicks; you can see kittiwakes and puffins on sea cliffs; you can watch badgers and bats.

And there are plenty more webcams all over the world to check out. Explore.org is a wonderful website with links to webcams globally. You can watch wolves, hummingbirds, African wildlife and much more; you can go underwater to see sharks and coral reefs. Links to these webcams tell you the best viewing hours, as of course there's no guarantee that anything will be happening while you visit.

But there's a special exhilaration in watching brown bears catching salmon in a waterfall in Alaska and knowing that it's happening in real time, while in Kenya two hippopotamuses are wallowing in a watering-hole. I could happily spend hours here, moving from one location to another. There's a huge range of films and photographs, too, so it's a great way to start learning more about extraordinary wildlife both far away and close to home.

“

WITH WEBCAMS YOU CAN WATCH WOLVES, HUMMINGBIRDS, AFRICAN WILDLIFE AND MUCH MORE; YOU CAN GO UNDERWATER TO SEE SHARKS AND CORAL REEFS.

BE A CITIZEN SCIENTIST!

There are various projects you can take part in which help to monitor wildlife and keep track of which species are thriving, and which are in decline. These citizen science programmes are open to anyone who wants to join in and contribute.

One of the best known of these projects in the UK is The Big Garden Birdwatch organized by the Royal Society for the Protection of Birds (RSPB) every year on the last weekend in January. To take part, you watch birds for an hour and record what you see - you can do this in your own garden, or in a park, nature reserve or out in the countryside. Results are submitted online and the findings are announced a few months later.

FrogWatch USA is a citizen science programme which will teach you how to listen to frog and toad calls in your area, and contribute to local research monitoring frog populations. You don't need to be an expert - local or online training is offered, and all you have to be is interested in frogs and keen to learn more!

iNaturalist has a wealth of information, helping you to identify plants and animals you see around you. By joining, and using its app, you'll be in touch with experts and enthusiasts around the world. One of their recent projects, for Australian citizens in the south and east of the country, is Environment Recovery Project, where you can help assess the impact of bushfires on wildlife, by sending photographs and information.

Most projects like this need no specialist knowledge or will give you the simple training you need. You don't even need to go out scouting: there are projects you can join from home, such as Project Plumage, Chimp&See and PELIcams among many projects organized by Zooniverse. All you need to take part in these is an internet link and some time to spare.

In the USA, Thinking Animals United has a list of citizen science projects to join, and National Geographic (worldwide) is another good place to look for citizen science projects you can join online.

Being part of any of these is a way of adding to the fund of knowledge about animal behaviour and habitats, so that we can help to protect wild animals around the world.

MAKING A WILDLIFE-FRIENDLY GARDEN

If you're lucky enough to have your own garden, you can make it a haven for wildlife. Even if there's not much space, you could have a small shrub with berries, plants for bees and butterflies, and a pond to attract dragonflies, damselflies and even frogs and newts, as well as providing drinking water for birds and small animals. There's plenty of advice on this online.

Many schools now have wildlife gardens too – if yours hasn't, you could suggest it to a teacher and maybe get together a group of interested people to make and look after it.

A super-tidy garden – with short-cut lawn, beds kept neat and weeded, dead-heading done regularly – isn't the best for wildlife! If you're not the gardener, you may need to do some persuading, but there are lots of smaller changes you could make too:

GARDENING FOR WILDLIFE

FEED GARDEN BIRDS

Feed garden birds from a bird table or from hanging feeders. Putting out seed, peanuts, fat balls, sunflower hearts and food scraps will bring birds regularly to your garden.

GROW BEE-FRIENDLY PLANTS

Grow bee-friendly plants like salvias, lavender, marigolds and sunflowers. It's fun to grow plants from seed, and many can be started on a windowsill indoors. Sunflowers are among the easiest – and they grow so fast that you can almost watch them shoot up!

KEEP A WILD LAWN

If there's a grass lawn, leave part (or all) of it unmown. Not only is it interesting to see what comes up, but flowering grasses are beautiful, and there may be wild flowers that attract insects.

If the grass is mown, ask whoever does it to let it grow longer between each mowing. Wild flowers are valuable to bees, and in very hot weather, grass left longer won't dry out as quickly as a short-mown lawn.

STOP DEAD-HEADING

When flowers have faded, don't cut them off – leaving them to set seed can provide autumn and winter food for birds.

PROVIDE A SHELTER

A small log-pile, or a heap of cut twigs, can provide shelter for frogs, toads or hedgehogs, as well as for beetles and spiders.

PLANT A TREE

If you have space, you could plant a tree or shrub (even a small one in a pot) with berries to feed birds in winter.

PROVIDE WATER

If you haven't got space for a pond, you could still have a water garden in a barrel, sink or washing-up bowl sunk into the ground. Be sure to place stones or a branch in the water so that if a creature accidentally falls in, it can climb out to safety.

If you don't have a pond, put out saucers or shallow bowls of water during hot, dry weather. Many animals, birds and insects struggle to find water in the heat, so this will help them.

MAKE OR BUY A 'BEE HOTEL'

Make or buy a 'bee hotel' to provide shelter for solitary bees and bring pollinators to your garden. Garden centres sell them but you could also make your own, from bamboo canes, pinecones, flower-pots or straw - check online for instructions.

PROVIDE A NESTING BOX

A nesting box can attract local birds to breed in your garden. Make sure a box is placed high enough to be safe from cats and ideally with the shelter of shrubs or a hedge nearby.

AVOID HARMFUL PRODUCTS

If someone else does the gardening, talk to them about avoiding harmful insecticides or slug pellets. Not only do these kill a range of insects - they can be dangerous to small mammals too.

HAVE A COMPOST BIN

Have a compost bin in which you put vegetable and fruit peelings, eggshells and other raw leftovers. When rotted, the compost will enrich the soil and provides a habitat for worms. Composting also reduces the amount your household puts into landfill.

COLLECT RAINWATER

If there isn't a water butt in your garden, can you find room for one? This will collect rainwater from the house roof or from a shed. Rainwater is better than tap water for topping up a pond - and it's a great way to save water, too.

USE POTS

If you haven't got much space, or if your garden is all paved over, you can still grow flowering plants in pots or a window box. You can find plenty of advice online about container gardening, helping you to choose plants to attract butterflies, bees and other insects.

ENJOY YOUR GARDEN!

Having your own little patch - whether it's a garden or a few pots or a window-box - can give you such pleasure: tending the plants, watching them grow to flower or seed, knowing that you're helping local wildlife by providing pollen or nectar, seeds or shelter.

NATURE AND MENTAL HEALTH

Being outside in natural surroundings is a great way to see and learn about wildlife and habitats. But there's another benefit – it's good for our mental health.

Being in a wood or park, or by a river or the sea, can take us away from our busy lives and worries and have a calming influence. Gardening, walking, exercising outside or watching birds or animals can reduce stress and anxiety and help give us a perspective beyond our own lives. Being by the sea, listening to the sound of waves as they rise and fall, can be almost mesmerizing.

During the COVID-19 pandemic many people took the chance to be outside whenever they could, listening to birdsong and watching the advance of the seasons. For those living in cities or dense suburbs, this was sometimes difficult – but parks, river banks and canal paths provided places to walk and play in the open air. Perhaps there was a special comfort in knowing that although our own lives were disrupted, birds were still nesting, trees coming into leaf and butterflies emerging as usual.

Mental health organizations now use the term ecotherapy, which they think is so important that they'd like to provide it for people suffering from depression. It seems that there's a deep-rooted need in us to be near plants, trees, water and open sky, and that we respond to their presence in ways that aren't obvious to us. Watching animals, birds and insects is rewarding in itself, and you may be doing yourself some good, too, by becoming mindful, observant and appreciative of small things.

“

WALKING, EXERCISING OUTSIDE OR WATCHING BIRDS OR ANIMALS CAN REDUCE STRESS AND ANXIETY.

PROTECT
AND
PROTEST

“

WHAT CAN YOU DO TO CHANGE THINGS FOR THE BETTER?

What should you do if you find an injured wild animal? Or a young bird that seems to be lost and alone? What if you see someone deliberately harming an animal?

Here's the advice from RSPCA in the UK, which is great to keep in mind. If you live elsewhere in the world, you will find similar advice from local welfare organizations. It's also a good idea to know in advance which emergency number to call, if needed.

- With an injured wild animal, watch first to see how badly hurt it is. It may recover on its own.
- Small animals such as rabbits and birds can be taken to a local vet. They don't usually charge for treating or euthanizing (putting to sleep) injured wild animals.
- Importantly, keep yourself safe. Be very careful when approaching or picking up wild animals. Badgers, for instance, can give a very painful bite.
- Be particularly careful if an animal is injured on or beside a road.
- Phone an animal welfare organization if it's a situation you can't deal with yourself or you're not sure what to do next.
- If you find a dog that seems to be lost, tell your local authority - they may have a dog warden who looks after stray dogs and tries to find their owners.
- Healthy stray cats aren't collected by local authorities, as many of them are feral (living wild). These cats won't usually let you come near. But a cat that's friendly and approachable is likely to belong to someone. If you're sure it's strayed from its home and you can safely catch it and put it in a box or basket, you can take it to a vet to find out if it's microchipped. If it is, its owner can be traced; if not, the vet may be able to care for it, or pass it on to a shelter nearby.
- Baby birds often appear to be lost when in fact their parents are nearby. In most cases you should leave

them where they are. Don't try to put a baby bird back in its nest (even if you know where the nest is) but if it's in danger, for instance in a road, you could move it to a more sheltered spot.

- If you think an abandoned or injured bird does need help, contact your local wildlife rehabilitator. Don't attempt to rear a fledgling bird yourself - they need very specialized care and are unlikely to survive without it.
- If you find an animal trapped and in danger, for instance a horse that's stuck in deep mud and is in distress, contact a local organization such as the RSPCA if you are quite certain that the animal can't free itself.
- If you see loose animals in a road - for example sheep or horses - and no one else seems to be aware, contact the local police.
- If you see deliberate cruelty, report it to local welfare organizations or the police. But be very careful not to put yourself in danger.

CAMPAIGNING FOR ANIMALS

What can you do to change things for the better, and encourage others to follow?

This book is about living by your principles, making animal awareness an important part of your life and avoiding cruelty in everything you do. By adapting your daily behaviour, you're showing others that this is important to you and how our habits can change. This in itself is a way of challenging patterns of buying and eating which many people don't think to question.

If you want to go further than that there are many things you can do.

There's a huge range of organizations you could join throughout the world - see the list at the end of this book. Some have separate sections for young people. You might decide to join one of them and focus on its campaigns and activities. This will certainly give you plenty to do - petitions to sign, letters to write, practical things to help wildlife or ways to raise money.

It can become overwhelming to belong to too many organizations, as well as expensive - but you can still follow on social media without needing to join.

YOU CAN HELP ANIMALS IN A RANGE OF WAYS:

Sign petitions. By following animal organizations on social media you will see plenty of these – you could even start your own.

Some shelters need temporary foster homes for rescued animals – maybe you could help out in that way if your home circumstances allow.

There may be ways to help animals in practical ways, for example by volunteering at a local animal shelter. Shelters often need helpers for dog-walking, looking after animals or even cat-cuddling to help rescued cats become socialized, ready for new homes.

Contact your local government representatives. You could write, for example, to state your support for a campaign or projects like wildlife-friendly gardening, and ask whether your representative will add their name too. There may even be a chance to go and meet them, and tell them your concerns. This is a good way of engaging in local democracy. Look online to find out who your local representatives are, and how to contact them.

If horses and ponies are your particular love, there may be a rescue home near you that would welcome your help with care, grooming and exercise.

There's probably a campaigning group near you for Greenpeace, World Wide Fund for Nature or other organizations. Joining would give you the chance to take part in fundraising and campaigning as well as keeping you up to date with local events.

Try to vary what you post on social media. If you post nothing but horror stories, it's likely to put people off rather than attract them to your cause. Post with a purpose – asking people to sign a petition, write a letter, or take a pledge, for example #plasticfreeJuly, #buynothingday or #veganuary. Post success stories or inspiring photographs of animals, insects and birds.

Have a birthday fundraiser on Facebook or other social media. Last year I raised money for Compassion in World Farming this way.

Nature reserves regularly organize volunteering projects to maintain their reserves. This is a great way of learning about wildlife and habitats.

Dramatic performances or 'installations' can be very effective at getting attention. Extinction Rebellion is particularly good at this, with its 'die-ins' to raise awareness of climate change. One recent XR action involved campaigners silently placing pairs of children's shoes, perfectly aligned, on Hastings pier on the south coast of England. This was a simple, poignant way of showing that lives are at risk because of climate breakdown.

Tell your local newspaper or news website what you're doing and why – they may send a reporter and photographer and run an article. If you're confident, you could ask for a slot on local radio to publicize your campaign. Local radio, newspapers and sites are always looking for stories and they're especially likely to be interested in an offer from a committed young person.

Practise being interviewed, ideally with a friend or a group. How will you get your point across to people on the street? How will you make your point briefly and effectively to a newspaper reporter?

Dressing up or wearing masks can be a good way to attract attention – campaigners against insecticides often dress up as bees.

If you can get a group of people together you could design placards or make leaflets and have an 'outreach' event in a town centre or shopping street. The focus of this could be to ask people to sign a petition, join a group or write to your local government official.

Use social media for your own campaigning, if you can. Retweet on Twitter, post on Instagram, use hashtags!

Build up a stock of simple craft materials for posters and placards. Ask your local supermarket if they have cardboard you could use. Big felt-tip pens, coloured paints and glue will be useful.

If you're confident at speaking, maybe you could ask to speak at school about your views on animals and how you want things to change.

STAYING MOTIVATED

Because I follow various animal welfare organizations and campaigns, I find regular emails in my inbox and alerts on social media. Some of these show terrible cruelty. Just today, I saw two stories that were distressing to look at and one that had a warning I chose not to go past.

Because I've been campaigning for years, I'm used to seeing evidence of horrors inflicted on animals - but even so, I'm regularly shocked and appalled. There are times when I feel like despairing. There seems no end to the ways in which animals are treated as mere objects; no end to the callous disregard for suffering.

It can be hard. Sickening. Gut-wrenching.

But are we going to give up, and accept it? I hope not.

So, importantly: we need to look after ourselves. Too much horror can rob us of all purpose, make us feel powerless and even affect our mental health.

What are the best ways to cope?
One is to try to focus on the positives - the areas where we can make a difference and begin to see a change for the better. In fact, showing people horrific photographs can have the effect of making them turn away, rather than engaging with the issue - you'll probably find this if you try to persuade others. People just don't want to know. And in the same way, even if you do want to know, making yourself feel miserable and powerless isn't the best motivation.

I take inspiration from high-profile campaigners, especially Earthling Ed, Chris Packham, Bella Lack, and John Oberg. For other inspiring people around the world, see the list at the end of the book - and maybe you can add more of your own? There are sure to be others making themselves known, especially as young people across the world add their voices and their passion. 'Passion' is an overused word in many contexts - but animal advocates really do feel passionate about their causes.

I'm encouraged, too, by the sheer number of organizations around the world; while writing this book I learned about more and more on top of those I already knew. It's impressive that there are so many people who care deeply about animals and how we treat them - people with immense knowledge or years of practical experience, people who inspire others, who organize campaigns, fundraisers and projects, people who devote time and energy to their causes. Collectively that's a huge number of voices speaking up for animals.

Will you add *your* voice?

Find ways to encourage birds, insects and wildlife in your own garden, balcony, window box or school grounds. This can be a great way to catch the interest of other people and to show them how creatures live and behave.

Try to find people who feel the same as you do. Their support will make you feel purposeful.

Join an organization or follow it on social media. That way you will hear of successes and small victories - these show that campaigning can work.

Make your campaigning creative by thinking of the best ways to persuade people. Use artwork, dressing up, and even humour to get your point across.

Take part in small local actions or campaigns – persuading a local clothes shop not to sell fur accessories, or asking your school to supply more vegetarian and vegan choices for lunch. Every small step helps.

Follow campaigners you admire on social media. They will give a lead you can follow, such as sharing posts, signing petitions or signing up to newsletters.

Find ways of being in the natural world as often as you can. Depending where you live, this may be a local park, or a river or canal path, or a wood or nature trail. There is always something of interest, and the more you look, the more you see.

Be kind to yourself. You can't change the world all on your own, and none of us is perfect. But we can take small steps that influence others. Take time to give yourself a break and do things that you know will give you a lift.

“

‘PASSION’ IS AN OVERUSED WORD IN MANY CONTEXTS – BUT ANIMAL ADVOCATES REALLY DO FEEL PASSIONATE ABOUT THEIR CAUSES.

HOW TO RESPOND TO THE THINGS PEOPLE SAY...

For some reason, there are people who like to try to catch out vegetarians and vegans. I'm not sure whether it's because we're seen as preachy, thinking of ourselves as better than everyone else - or could it be that those questioners secretly feel that we have a point, but they don't want to give up meat themselves, so try to belittle us instead?

Questions like *"Ah, but do you wear leather shoes?"* or *"Do you feed meat to your cat?"* give the feeling that people want to demolish our position by making us seem like we are being hypocritical.

My answer (apart from *"No, I don't wear leather shoes except for my ancient gardening boots"*, and *"Yes, I do feed meat to my cats"*) is that I'm not attempting to hold myself up as a model of the perfect vegan lifestyle. I'm aware of my own inconsistencies. But I've decided on certain principles to live by and I do the best I can. Who can argue with that?

Whatever your position, don't let anyone make you feel ashamed of it - one sure thing is that you're doing better than those who have no principles other than eating and buying whatever they like.

Here are a few comments and questions you might come up against, and my thoughts about them. Sometimes it can help to think about your answers in advance so you're ready to state your case confidently.

"What if you're on a desert island with a pig, and you'll starve unless you eat the pig – what will you do then?"

Oddly, this is a question often asked. My answer is that I'd watch to see what plants, seeds, etc the pig was eating, and eat some of that - but of course it's not really a serious question. The question can be turned around: "What if you're in a world of nearly 8 billion people, all of whom need feeding? Will you choose to eat fruit, vegetables, seeds, grains and pulses, or will you choose to eat dead animals even though the planet can't afford that?"

"But if everyone stopped eating meat, those animals wouldn't have a life at all."

My answer to that one is that I'd prefer animals not to be bred solely for intensive farming systems or to be slaughtered well before they reach maturity. The other side of this is -

"If we didn't eat animals, they'd take over the world."

People often say this, but it doesn't really make sense. For example, there are an estimated 50 billion chickens in the world, most bred for intensive farming and slaughter. If we didn't want to eat chicken, these birds wouldn't exist - it's as simple as that. And are these people really worried about animals taking over, or are they just trying to 'catch you out'?

"So – what do you eat?"

This shows a distinct lack of imagination - as if, when you take meat off your plate, there's nothing left! Plant-based foods include rice, pasta, bread, countless fruits and vegetables, seeds, grains, nuts and the infinite variety of things that can be made by combining them.

"Isn't it boring eating nothing but plants?"

No. Isn't it dull eating dead animals day after day?

"But where do you get your protein?"

When does anyone ever worry about anyone else's protein intake other than when learning that they're vegetarian or vegan? Racehorses, elephants, hippopotamuses, giant pandas and giraffes don't eat meat, but they seem to get by. An increasing number of top athletes are vegan too. The serious answer is that our protein can come from beans, seeds and grains - and that many people eat far more protein than they need anyway.

"Is it expensive to buy vegetarian or vegan food?"

It could be, if you always bought ready meals and didn't cook from ingredients - but the same would apply to a meat-based diet, too. Some things are more expensive, for instance plant milk usually costs more than dairy milk, but I'd say that overall a plant-based diet can often cost considerably less than a meaty one.

"Don't plants have feelings too?"

If your questioner seriously believes this and cares about it (they rarely do), they'll be fruitarian and eat only seeds, fruits, berries, beans and nuts - i.e. not the plants themselves. Meat-eaters don't seem to realize that they're indirectly eating more plants than vegetarians and vegans are - because the animal they're eating has depended on plant food.

"But we're meant to eat meat. We have canine teeth."

Our teeth - incisors, canines and molars - suggest that we are omnivores, while our digestive systems are like those of apes and monkeys. Their diet consists of nuts, fruits, seeds, insects and - for some - occasional flesh. Not heavily meat-based, then. And unlike apes and monkeys, we can make ethical decisions about what we eat and why.

"Are you allowed to eat fish / seafood / chicken?"

Questions like this imply that there's a list of rules imposed on us by someone else. My answer would be that I make my own rules and choose not to eat anything that's had to be killed. Nothing with a face for me!

"But do you cheat sometimes?"

Who would I be cheating? Only myself - so no, I don't. If as a vegetarian or vegan you mistakenly eat something containing meat, it might feel deeply unpleasant and you wouldn't choose to do it, but you shouldn't feel guilty.

"I love animals, I really do, but I couldn't give up meat."

Hmmm. To that person I would suggest that they could at least make a commitment to eating less meat.

"People are suffering all over the world. Why don't you campaign for people, rather than for animals?"

This is an odd one, with its assumption that we can commit only to one cause. In my experience, most people who care about animals also care about human suffering, and want fair treatment for all living things.

"Do you mind if I eat meat in front of you?"

Occasionally someone asks this, and I appreciate the thought. Several of my friends eat meat, as does everyone in my family other than me. Some strict vegans refuse to sit at a table where anyone's eating meat, but I don't go that far, even though I don't like it. I have to accept that I'm not responsible for other people's choices - only for my own.

Generally, I only get into conversations like those above when they're started by someone else, or if someone is genuinely interested in my reasons for being vegan. My hope is that - with more and more pressing reasons for eating less meat or no meat at all - more people will decide for themselves to give it up, and to pursue that idea into living as sustainably and kindly as they can.

OUTRO

I hope that this book has been helpful, stimulating and maybe challenging. I hope it has helped you to make some choices about how to live a kinder and more cruelty-free life. It may also have helped you to decide how you'd like to live your life in the future.

At first it may seem difficult to stick to the rules you make for yourself. But soon these choices will become so much a part of you that there's no need to struggle every day. Rather than missing the things you can't have (or rather, that you've decided not to have, because no one but you can make that decision), such as meat or fast fashion, look at all the new possibilities opening up.

If you feel strongly about cruelty and the exploitation of animals, you can at least know that you're not part of it. By making your life as cruelty-free as you can, you're showing that it's possible. We don't have to buy and consume everything that's offered to us, regardless of the cost to animals and the environment.

We can make better choices.

I've learned a lot while writing and researching this book. When I first thought of writing it, my title was *Live Kindly, Tread Lightly*. That's what I aim to do, and if more of us try to live without causing harm, the world will be a kinder, greener, more sustainable place - for animals, for habitats, for all of us.

Thank you for reading! Now, over to you.

“WE CAN MAKE BETTER CHOICES.”

“

IF MORE OF US TRY TO LIVE WITHOUT CAUSING HARM, THE WORLD WILL BE A KINDER, GREENER, MORE SUSTAINABLE PLACE.

ORGANIZATIONS AND USEFUL WEBSITES

This is a long list, covering a range of organizations and special interests worldwide. I hope that wherever you are in the world you'll find it useful for discovering organizations to join and support, or for projects and research, widening your knowledge and keeping up to date.

American Kennel Club, www.akc.org
Includes advice on how to choose a puppy and all aspects of care.

Animal Aid, www.animalaid.org.uk
One of the world's oldest established animal rights groups, which campaigns against animal abuse and encourages cruelty-free lifestyles.

Animal Aid USA, www.animalaidusa.org
A rescue organization devoted to rehoming dogs through a large network of volunteers.

Animal Defenders International (ADI), www.ad-international.org
includes ADI America, ADI Europe, ADI UK and ADI South America. Campaigns on a range of issues including animals in entertainment, research and farming.

Animal Rebellion, www.animalrebellion.org
Part of Extinction Rebellion with local groups across the world. Campaigns for a humane and sustainable plant-based food system to stop mass extinction, alleviate climate breakdown and ensure justice for animals.

Association of Zoos and Aquariums, www.aza.org
US organization dedicated to the advancement of zoos and aquariums in conservation, education, science, and recreation. (See also **WAZA**)

Australian National Kennel Council, www.ankc.org.au
Provides information and advice on pure-breed dogs and responsible ownership in Australia.

Birdlife International, www.birdlife.org
A global network of conservation groups working to protect birds and their habitats, with a focus on sustainability and preserving biodiversity. Includes **Birdlife Australia**, **Forest and Bird New Zealand**, **Birdlife Americas**, and **Birdlife Europe and Central Asia**.

Born Free, www.bornfree.org.uk
Wildlife charity which protects threatened or endangered animals, conserves habitats and ecosystems, and opposes the exploitation of wild animals in captivity.

Buglife UK, www.buglife.org.uk
Aiming to save insect life. Information, identification guides and various projects including wildlife gardening advice.

Butterfly Conservation, www.butterfly-conservation.org
Aims to halt and reverse the declines of butterflies and moths in the UK. Organizer of The Big Butterfly Count citizen science project in late July and August each year.

Compassion in World Farming, www.ciwf.org.uk
Campaigns peacefully to end factory farming.

As a Compassion in World Farming Visionary, I support the organization's vision:

WE SEEK FOOD AND FARMING POLICIES THAT PROMOTE:

- Good health by ensuring universal access to sufficient and nutritious food.
- Sustainable farming methods, which support rural livelihoods and relieve poverty.
- Protection for the planet and its precious resources: soil, water, forest and biodiversity.
- Reduced emissions of greenhouse gases and other pollutants from agriculture.
- Humane farming methods which promote the health and natural behaviour of sentient animals and avoid causing them pain and suffering.
- Reduced consumption of animal products in high-consuming populations to meet environmental, health and sustainability goals.

Conservation International, www.conservation.org
Promotes responsible and sustainable care for nature and biodiversity across the world.

Cruelty Free International, www.crueltyfreeinternational.org
Works to end animal experiments around the world. Approves cruelty-free products with the Leaping Bunny symbol.

Cruelty-Free Kitty, www.crueltyfreekitty.com
A database to help you stop buying animal-tested products.

Earth Overshoot Day, www.overshootday.org
This organisation marks the day each year by which humanity's demand for food, energy and resources goes beyond what the Earth can provide. It aims to #MoveTheDate by encouraging us to live more sustainably.

Earthwatch, www.earthwatch.org
This international environmental organization operates across six continents to connect researchers with citizen scientists, and has offices in the US, UK, Australia, Japan, and India.

The Ethical Consumer, www.ethicalconsumer.org
A UK-based rating system to help you shop ethically.

Explore.org, www.explore.org
Films, photographs and webcams from every continent.

Extinction Rebellion, www.rebellion.earth
An international movement that uses non-violent civil disobedience to try to halt mass extinction and minimize the risk of social collapse. It focuses on political change over personal change.

Farms Not Factories, www.farmsnotfactories.org
Focuses on pig farming. Publicizes the damage caused by factory pig farming to animals, human health and the environment, and urges consumers to only buy local, high welfare and ethically produced pork.

Fauna and Flora International, www.fauna-flora.org
The world's oldest international conservation organization. Its mission is to conserve threatened species and ecosystems worldwide.

Freedom for Animals, www.freedomforanimals.org.uk
One of the UK's longest-running animal charities whose work focuses on issues affecting creatures held captive in circuses, zoos and aquariums, as well as those used in TV and film, live animal displays and the exotic pet trade.

Friends of the Earth, www.foei.org
Founded in the US, now an international grassroots network of environmental organizations in 74 countries.

Global Animal, www.globalanimal.org
A California-based online news magazine and social community for all things animal.

Global Animal Partnership (GAP), www.globalanimalpartnership.org
Texas-based, operating in seven countries to provide welfare certification at all stages of meat and dairy production.

Good On You, www.goodonyou.eco
News and information on sustainable and ethical fashion with brand ratings.

Greenpeace, www.greenpeace.org
This organization uses non-violent creative action to pave the way towards a greener, more peaceful world, and to confront the systems that threaten our environment. Links to sections for **Australia, New Zealand, the UK, USA** and many other countries.

Happy Cow, www.happycow.net
Browse the site or download the app to find vegetarian and vegan eating places nearby, wherever you are.

Hawk Conservancy Trust, www.hawk-conservancy.org
Works for the conservation of birds of prey. The Trust manages the National Birds of Prey Hospital, where injured or orphaned birds are treated before being released into the wild.

The Humane League, www.thehumaneleague.org
A global movement to protect animals, based in the US.

Humane Slaughter Association, hsa.org.uk
A UK charity, respected internationally, working to improve food animals' welfare.

Humane Society of the United States,
www.humanesociety.org
Driving impactful change to end suffering for animals.

iNaturalist, www.inaturalist.org
A nature app you can take with you everywhere on your phone, it helps you identify the plants and animals around you.

International Anti-Poaching Foundation, www.iapf.org
Supporting community-led conservation to end poaching in Africa, alongside education and employment opportunities.

International Fund for Animal Welfare (IFAW), www.ifaw.org
A global non-profit organization working in more than 40 countries around the world to rescue animals and to restore and protect natural habitats.

IUCN (International Union for Conservation of Nature), www.iucn.org
The global authority on every aspect of the state of nature and sustainable development. Its separate website IUCN Red List, www.iucn.redlist.org, is the world's most comprehensive information source on the worldwide extinction risk status of animal, fungus and plant species.

The Jane Goodall Institute, www.janegoodall.org.uk
A global organization that aims to help people to make a difference for all living things. Dame Jane Goodall's Roots & Shoots is a network and education programme for young people.

Jewish Vegetarian Society, www.jvs.org.uk
Hosts events, workshops and cooking demos.

The League Against Cruel Sports, www.league.org.uk:
Founded in 1924 with the aim of banning animal hunting in the UK. Still fighting to protect foxes, hare and deer from illegal hunting.

Live Kindly, www.livekindly.co
News, features and shopping guidance on how to live kindly.

The Mammal Society, www.mammal.org.uk
Surveying the UK's mammal population. Provides the Mammal Mapper app for recording your own surveys.

Marine Conservation Society, www.mcsuk.org
The UK's leading marine charity with campaigns from responsible seafood to microfiber pollution.

Meat Free Mondays, www.meatfreemondays.com
Aims to raise awareness of the environmental impact of animal agriculture and industrial fishing. Recipes, news and information on reducing meat consumption.

Meatless Monday,
www.mondaycampaigns.org/meatless-monday
US-based global movement. Plant-based recipes, resources and information.

National Geographic, www.nationalgeographic.com
In-depth features with great photography and infographics.

Natural Resources Defense Council, www.nrdc.org
US-based international organization combining members, scientists and lawyers.

Naturewatch Foundation, www.naturewatch.org
Its mission is to end animal cruelty and improve animal welfare standards around the world. Produces its own Compassionate Shopping Guide and has its own cruelty-free mark for products guaranteed not to have been tested on animals at any stage. See this online at www.compassionateshoppingguide.org

Oceana, www.oceana.org
Aims to protect and restore the world's oceans. Become a "wavemaker" with their regional sites and groups.

One Green Planet, www.onegreenplanet.org
An online guide to making choices that help people, animals and the planet. US-based, focusing on cruelty-

free food choices as well as a range of welfare and environmental issues.

PETA (People for the Ethical Treatment of Animals), www.peta.org
Against all animal exploitation.

Rainforest Action Network, www.ran.org
A US organization with a focus on forests, climate and human rights.

RSPB (Royal Society for the Protection of Birds), www.rspb.org.uk
The largest conservation charity in the UK. There are resources for families and children including projects, games and film clips.

RSPCA (Royal Society for the Prevention of Cruelty to Animals), www.rspca.org.uk
Works to prevent animal pain and suffering. Young RSPCA is for people of 18 and under and has its own website, young.rspca.org.uk. RSPCA Australia, www.rspca.org.au; SPCA New Zealand, www.spca.nz; American SPCA, www.aspca.org

SAFE for Animals, www.safe.org.nz
New Zealand's leading animal rights charity.

Stop Circus Suffering, www.stopcircussuffering.com
A global campaign to end the suffering of animals in circuses. Part of Animal Defenders International, running awareness campaigns, lobbying governments and conducting investigations.

Stop Ecocide, www.stopecocide.earth
A global campaign aiming to change international law.

Surge, www.surgeactivism.org
Promoting vegan education and community activism.

TED, ideas worth spreading, www.ted.com
A US-based organization with global reach. TED stands for Technology, Entertainment and Design, and the 'TED talks' cover a huge range of subjects. Search "animal welfare" for many fascinating talks.

Thinking Animals United, www.thinkinganimalsunited.org
New York based, with many international partnerships.

UK Student Climate Network, ukscn.org
Youth strikers with over 100 local groups across England and Wales.

Vegan Muslim Initiative, www.veganmuslims.com
Information, recipes and resources.

The Vegan Society, www.vegansociety.com
Advice on living a vegan lifestyle, and home of the recognizable Vegan Trademark.

The Vegetarian Society, www.vegsoc.org
Information on vegetarian food. Also **Australian Vegetarian Society**, veg-soc.org.au; **New Zealand Vegetarian Society**, www.vegetarian.org.nz, **North American Vegetarian Society (NAVS)**, www.navs-online.org

Viva!, viva.org.uk
Campaigns, information and local directories on veganism.

World Association of Zoos and Aquaria, www.waza.org
Its goal is to guide, encourage and support zoos, aquariums in animal care and welfare, environmental education and global conservation. Includes British and Irish Association

of Zoos and Aquaria (BIAZA), www.biaza.org.uk, European Association of Zoos and Aquaria (EAZA), www.eaza.net

Whale and Dolphin Conservation, www.uk.whales.org
The leading UK charity dedicated to the protection of whales and dolphins. Aims to end captivity, stop whaling and create healthy seas.

Wildlife Conservation Society, www.wcs.org
US conservation charity based at Bronx Zoo.

World Animal Protection, www.worldanimalprotection.org.uk
A global organization with millions of supporters worldwide working to end needless suffering of animals.

WWF (World Wide Fund for Nature), www.wwf.org.uk
The world's leading independent conservation organization. There's also a youth section to support and empower young people.

Wild Justice, www.wildjustice.org.uk
Fights wildlife crime on behalf of threatened and endangered animals.

The Wildlife Trusts, www.wildlifetrusts.org
Made up of 46 trusts around the UK.

Woodland Trust, www.woodlandtrust.org.uk
Aims to reverse the decline in native woods and trees. It plants trees, restores ancient woodland and fights climate change in the UK.

Zoo and Aquarium Association (ZAA),
www.zooaquarium.org.au
ZAA and members lead over 100 breeding programs, all working together to improve conservation and education globally.

Zooniverse, www.zooniverse.org
Online citizen science projects on a range of topics.

ZSL (Zoological Society of London), www.zsl.org
An international conservation society that runs two zoos in the UK: London and Whipsnade.

INSPIRATIONAL ANIMAL, WILDLIFE AND ENVIRONMENTAL ADVOCATES ON SOCIAL MEDIA:

This is just a selection, and no doubt you'll find more along your cruelty-free journey. Several of these people have ambassador roles with wildlife or conservation organizations. Many organizations are looking for young people to be wildlife ambassadors for the future - could you be one? Look out for opportunities!

Rachel Ama, British vegan cook, blogger and food writer. Video demonstrations and recipes at www.rachelama.com Instagram and Twitter: @rachelama_

Arjun Anand, Indian photographer who travels the world, photographing exotic wildlife. www.arjunanand.com Instagram: @arjunanandphoto

Liz Bonnin, French-Irish biochemist, wild animal biologist and presenter; president of The Wildlife Trusts, UK. www.lizbonnin.com Instagram and Twitter: @lizbonnin

Genesis Butler, young North American animal rights activist and vegan. The youngest person ever to give a TED talk, at the age of 10. Instagram: @genesisbutler_ and Twitter: @genesisbutlerv

Mya-Rose Craig (Birdgirl), young British Bangladeshi birdwatcher, fighting racism and campaigning for equality in nature and environment. Ambassador for Survival International. Instagram and Twitter: @birdgirluk

Earthling Ed (Ed Winters), British animal advocate and public speaker based in London. earthlinged.org Instagram: @earthlinged

Dame Jane Goodall, naturalist and conservationist with a special interest in chimpanzees; campaigns on a range of climate, environmental and welfare issues. Janegoodall.org Instagram and Twitter: @janegoodallinst

Kabir Kaul, conservationist and wildlife writer living in London. RSPB Youth Councillor and London National Park City champion. Twitter: @Kaulofthewilduk

Bella Lack, British conservationist and ambassador for the Born Free Foundation, SaveTheAsianElephants, RSPCA and Jane Goodall Institute. Instagram and Twitter: @BellaLack

Tim Laman, award-winning North American photographer and film-maker specializing in rare and endangered wildlife. www.timlaman.com Instagram: @timlaman

David Lindo, British wildlife presenter, writer and guide. Ambassador for @WildLondon and co-curator of London Wildlife Festival. www.theurbanbirderworld.com Instagram and Twitter: @urbanbirder

Dara McAnulty, naturalist living in Northern Ireland, author of Diary of a Young Naturalist. Instagram: @dara_mcanulty and Twitter: @naturalistDara

Megan McCubbin, British zoologist, photographer, conservationist and wildlife television presenter. Instagram: @meganmccubbin_photo and Twitter: @MeganMcCubbin

Macken Murphy, writer and vegan based in Boston, Massachusetts. Host of Species podcast, each episode an in-depth exploration of an animal species. Twitter: @SpeciesPodcast

John Oberg, an animal advocate from Washington D.C. Instagram and Twitter: @JohnOberg

Chris Packham, British naturalist, author, photographer and broadcaster, presenter of BBC Springwatch. www.chrispackham.co.uk Instagram: @chrisgpackham2 and Twitter: @ChrisGPackham

Greta Thunberg, Swedish climate and environmental activist. Instagram and Twitter: @GretaThunberg

Edin Whitehead, seabird scientist and conservation photographer from Auckland, New Zealand. www.edinz.com Instagram: @edinzphoto and Twitter: @edinatw

VEGETARIAN AND VEGAN RECIPES ON LINE:

There's a huge range available worldwide, including many by enthusiastic bloggers and YouTubers, so this is just a small selection of places to start. Have a browse and start cooking!

American Vegan Society, www.americanvegan.org

BBC Good Food, www.bbcgoodfood.com

Bosh!, www.bosh.tv

Delicious, www.deliciousmagazine.co.uk

Good Food Australia, www.goodfood.com.au

Meat Free Monday, www.meatfreemondays.com

Meatless Mondays, www.mondaycampaigns.org/meatless-monday

Vegan Australia, www.veganaustralia.org.au

Veganuary, www.veganuary.com

The Vegan Society, www.vegansociety.com

The Vegetarian Society, vegsoc.org

Vegetarians New Zealand, www.vegetarians.co.nz

Yummly, www.yummly.co.uk

THANK YOU

Thanks to my agent, Catherine Clarke, for suggesting I write non-fiction; to my friend and mentor Jon Appleton, for unfailing support; to booksellers extraordinaire Marilyn Brocklehurst and Tamsin Rosewell for help and encouragement; to Neil Dunnicliffe and the team at Pavilion for making it happen, and to Josephine Skapare and Sarah Crookes for the striking cover, illustrations and design; and to all those campaigners, activists and dedicated specialists who work so hard to make the world a better place for animals.